THE ICE AGE AND CLIMATE CHANGE

A Creation Perspective

THE ICE AGE AND CLIMATE CHANGE

A Creation Perspective

Jake Hebert

Dallas, Texas
ICR.org

THE ICE AGE AND CLIMATE CHANGE

A Creation Perspective

by Dr. Jake Hebert

First printing: February 2021

Series concept and direction: Jayme Durant, ICR Director of Communications
Senior Editor: Beth Mull
Editors: Truett Billups, Michael Stamp, Christy Hardy
Graphic Designer: Susan Windsor
Cover image: Vatnajokull glacier in southeast Iceland, BigstockPhoto

Unless otherwise indicated, Scripture quotations are from the New King James Version.

ISBN: 978-1-946246-34-9
Library of Congress Control Number: 2020951811

Please visit our website for other books and resources: ICR.org

Printed in the United States of America.

TABLE OF CONTENTS

FOREWORD

What would it have been like to be alive 5,000 years ago? Or 4,000 years ago? Would Earth have been warm and pleasant or cold and covered with ice? Would you have wandered as a nomad or lived in a cave?

Most of us don't worry about what life in the past was like. We live in comfortable homes and have food enough not to worry about our next meal. But there are some who think about such questions.

They believe that climate and food sources may be changing. Some scientists who study climate change and population growth suggest that too many people and warming temperatures could lead to starvation and human migrations.

Many alarmists are calling for a Green New Deal and population control. Their calls for action are said to be based on science. But there's great controversy over whether their science is valid, even among the conventional scientific community.

One group in particular questions the basis of these concerns. A portion of the Christian community believes the Bible is an important source of information for understanding Earth history and that it bears on this issue.

They believe the Bible reveals that Earth is only thousands of years old, people were created fully human only several thousand years ago, and a global flood changed the atmosphere, ocean, and Earth in major ways.

The Ice Age and Climate Change: A Creation Perspective is the most thoroughly researched, authoritative, and biblical exposé on this subject available today. It's written in a conversational, non-condescending manner.

Dr. Jake Hebert fully addresses the weaknesses of the conventional views and explains scientifically and biblically where climate hysteria may have gone off the rails. And he tells you upfront why this topic is important.

I wish I had written this book. But the author's writing is more elegant than mine and so well-documented that I believe it will become a classic. Jake, may the Lord bless you and the wind be always at your back.

Larry Vardiman, Ph.D.
Retired Senior Research Scientist, Astro/Geophysics
Institute for Creation Research

ACKNOWLEDGMENTS

I would like to acknowledge and thank the scientists who reviewed my work for accuracy, including Russell Humphreys, Timothy Clarey, Michael Oard, and Randy Guliuzza. (Any possible errors are, of course, my own.) I would also like to thank the editorial staff at the Institute for Creation Research for their editorial assistance and suggestions, and graphic designer Susan Windsor for her layout and design work. Finally, I would like to thank the ICR Board and ICR's late CEO, Dr. Henry M. Morris III, for authorizing this project, as well as our supporters whose generous donations made it possible.

1 Introduction

Summary: In recent years, Christians have become more alarmed over climate change as both secular and Christian voices urged them to heed the "all-but-unanimous scientific consensus" on the issue. Alarmists are correct about at least one thing: Those who reject evolutionary theory tend to be more skeptical of climate change. Conversely, those who accept evolutionary theory tend to be less skeptical of climate change. Why is this the case? This book will explore the connection between these tendencies and argue that Christians have good biblical and scientific reasons to reject climate change alarmism.

In June 2005, Andy Crouch wrote a *Christianity Today* editorial that began:

> The theory is taken for granted by nearly every scientist working in the field. But because it is difficult to confirm experimentally, a few vocal skeptics continue to raise pointed questions. The skeptics find a ready audience among evangelical Christians, with groups like Focus on the Family saying that "significant disagreement exists within the scientific community regarding the validity of this theory."
>
> I'm not talking about evolution. Or maybe I am.[1]

Instead, Crouch was actually describing the global warming issue, the claim that the earth is warming due to human-produced carbon dioxide emissions and that this warming could result in catastrophic climate change. Crouch thought evangelicals were being unreasonably skeptical on this issue. He urged them to heed the "all-but-unanimous scientific consensus" that global warming is real and requires a response.

Crouch is probably pleased that in the years since he penned that essay Christians have become less skeptical about the climate change claim. Pope Francis has called for

action to fight climate change.[2] Likewise, dozens of evangelical leaders signed the document "Climate Change: An Evangelical Call to Action," arguing that climate change is real and action must be taken.[3]

Just as Crouch thinks skepticism of global warming is unwarranted, he also thinks skepticism of evolution (and its accompanying billions of years) is unwarranted. He obviously thinks that the first chapters of Genesis can't be taken at face value. This is indicated by his claim that in response to evolutionary ideas "some Christians resorted to a wooden interpretation of the first pages of Genesis that was no better as science than evolution was as a worldview." Of course, what Crouch is calling a "wooden" interpretation of Genesis is really just a face-value reading of the text held by orthodox Christians for thousands of years. Clearly, Crouch accepts the conclusions of evolutionary theory, although he would probably try to argue that God somehow directed the evolutionary process.

Crouch is not alone. Dr. Katharine Hayhoe is an evangelical Christian, as well as an atmospheric scientist at Texas Tech University, and she is convinced that global warming is real and is a genuine threat to humanity. Numerous media outlets have profiled her, including a PBS NOVA documentary on climate change. In that documentary, she makes this statement about current carbon dioxide levels:

> We are so far outside the range of natural variability. We have not seen carbon dioxide levels like this in the history of human civilization on this planet.[4]

Dr. Hayhoe clearly does not hold to a recent creation. According to the Bible, human civilization is almost as old as the earth itself, since God created humans on Day 6 of the creation week. Yet, Dr. Hayhoe speaks of a time long before human civilization. Of course, a long period of prehistory assumes the millions of years of evolutionary belief.

> "Alarmists are correct about at least one thing: Those who reject evolutionary theory tend to be more skeptical of climate change."

I believe Crouch is correct that a connection does exist between Christian skepticism of global warming and skepticism of evolutionary dogma. Although there are obviously exceptions, I think it is safe to say that most Christians who doubt evolutionary claims also doubt that global warming is a serious issue, if they think the earth is warming at all. Likewise, those who accept some or all of the tenets of evolutionary theory tend to be more receptive to global warming claims.

So, what is the connection between evolution and global warming? In this book, I argue that much, perhaps most, climate alarmism is ultimately driven by a denial of the literal history recorded in Genesis. While not everyone who accepts evolution or an old earth is a climate alarmist (some evolutionists are outspoken critics of climate alarmism), it is also true that nearly every climate alarmist accepts at least some of the claims of evolutionary theory, such as the doctrine of an old earth. I don't think this is a coincidence. Climate alarmism, and environmental alarmism in general, are logical consequences of the evolutionary worldview.

For this reason, most of this book is an attempt to persuade you, the reader, that the Genesis record is true literal history. I seek to demonstrate that the Bible makes much better sense of the historical and geological big picture than the evolutionary story does. It also makes much better sense of past climate change, including the Ice Age. Although I do not intend this to be a technical book, I have included footnotes for those who would like to study this subject in more depth. You can read many of the cited references freely online.

I am a physicist, not a meteorologist or climatologist, but I have done serious research in several areas that are extremely relevant to the climate change issue. Some of this was my Ph.D. work on a possible connection between weather and climate, solar activity, and cosmic rays (discussed in appendix A). Some of it was a continuation of ice-core research by other creation researchers. Some of it was original research into the supposed evidence for the dominant Ice Age theory, the so-called *Milankovitch*, or *astronomical*, theory.

Although I have tried to keep an open mind as to whether or not global warming is occurring (right now I doubt it), I have always been skeptical of alarmism on this issue. This skepticism comes from my own research, as well as the frankly suspicious

behavior exhibited by high-profile alarmist scientists. However, this book will not focus on well-known scandals such as Michael Mann's hockey stick graph or the 2009 East Anglia University climategate email hacks.[5]

Instead, I argue that Christians seeking to properly understand and respond to the climate change issue must take into account the literal history recorded in Genesis 1–11. In the process, we will tackle what skeptics consider some of their strongest arguments against biblical creation. We'll address such questions as:

- Are deep ice cores irrefutable proof of an old earth?
- How can any rational person doubt that the earth is old when so many different dating methods agree with each other?
- Do secular scientists understand the cause of Ice Ages?
- Is there strong evidence for multiple Ice Ages over millions of years?
- Do plate tectonics disprove biblical creation?

In addition, we'll also discuss:

- The climategate scandal you haven't heard about.
- How the Bible explains the Ice Age.
- Why a large Ice Age woolly mammoth population in Siberia is a mystery to secular scientists but is explained by the biblical Ice Age model.

Obviously, if the Bible does a better job of explaining past climate change, then it clearly has relevance for future climate change. After showing that the Bible makes better sense of the scientific data, we will examine ways that evolutionary beliefs contribute to climate alarmism. Some of these ways are obvious (i.e., a denial of God's sovereign care over our planet), but some of them are much more subtle, leading climate researchers to misinterpret past climate data, including chemical "wiggles" in deep-sea sediments and ice cores.

The bottom line is that ignoring, denying, or downplaying the Genesis record leads to incorrect conclusions. What one believes about the past influences what one believes about the future. The history recorded in Genesis has much more relevance to this issue than most Christians assume.

"Obviously, if the Bible does a better job of explaining past climate change, then it clearly has relevance for future climate change."

References

1. Crouch, A. Environmental Wager: Why evangelicals are—but shouldn't be—cool toward global warming. *Christianity Today*. Posted on christianitytoday.com June 29, 2005, accessed November 7, 2016.
2. Richardson, B. Pope Francis calls climate change a 'sin.' *The Washington Times*. Posted on washingtontimes.com September 1, 2016, accessed April 22, 2019.
3. Climate Change: An Evangelical Call to Action. The Evangelical Climate Initiative. Posted on npr.org, accessed April 22, 2019.
4. Hayhoe, K. Quoted in *Decoding the Weather Machine*. PBS NOVA. Aired April 18, 2018, transcript accessed October 8, 2018.
5. For information on these scandals, see Hebert, J. 2019. *The Climate Change Conflict: Keeping Cool Over Global Warming.* Dallas, TX: Institute for Creation Research.

2 Why You Should Care About This Issue

Summary: Climate change alarmists have suggested taking extreme measures to fight global warming. These measures include genetic engineering of people to reduce their carbon footprints, government policies limiting the number of children a family can have, legally prosecuting climate change skeptics, and even killing much of the human population! As draconian as these proposals are, some people are convinced they're necessary to stave off a climate catastrophe. With these kinds of extreme measures being suggested, it isn't really possible to be indifferent about the issue of climate change. The subject deserves careful examination of the relevant biblical and scientific evidence.

When it comes to global warming, much of the environmentalist movement seems to be descending into madness. Activists are so convinced of an impending climate catastrophe that they have proposed radical measures to "save the planet." In 2013, the prestigious science journal *Nature* ran an article suggesting that a large portion of 77 million acres of uncultivated cropland in Eastern Europe should remain unused.[1,2] The rationale was that using the land for agriculture would, on balance, contribute to increased atmospheric carbon dioxide. Hence, the land should be left alone to fight global warming. When one thinks about how many hungry and undernourished people there are in the world, it seems bizarre that academics would be thinking of deliberately reducing food production, but that is indeed the case.

In 2014, New York University philosophy professor Matthew Liao suggested that society should engineer humans with characteristics to reduce our carbon dioxide emissions.[3] He suggested giving people drugs to take away a taste for meat, since cattle are

"In the spring of 2006, University of Texas herpetologist Eric Pianka created a stir at the Texas Academy of Science meeting by suggesting that it would be a good thing if 90% of the human race died!"

major greenhouse gas contributors due to their methane-containing flatulence as well as other greenhouse gases associated with agriculture.[4] He also suggested making people shorter, since smaller bodies require less energy to sustain than larger bodies. He also discussed the possibility of giving humans cat-like eyes to enable us to see better in the dark, thereby reducing our energy consumption and carbon dioxide emissions. Although Liao insisted these measures should be voluntary, he spoke approvingly of government policies that limit the number of children a family can have.[5]

Other environmentalists are even more extreme. In the spring of 2006, University of Texas herpetologist Eric Pianka created a stir at the Texas Academy of Science meeting by suggesting that it would be a good thing if 90% of the human race died! Pianka (dubbed "Dr. Doom" by his critics) stated that an airborne version of the Ebola virus would be a good way for that to happen. Even more disturbing, eyewitnesses reported that the teachers in attendance gave him an enthusiastic standing ovation.[6-8] Sad to say, I've gotten so used to lunacy in academic circles that this might not have

made much of an impression on me. However, I was shocked to learn that Pianka made these remarks at Lamar University, the very school where I was currently working. Incredibly, even though this news story was blowing up the internet and blogosphere, there was no coverage of the incident in our local media.[6,8]

I distinctly remember Lamar University hosting that particular Texas Academy of Science meeting. I was not feeling well that day. I went home as soon as I finished teaching my morning classes so I did not actually attend Pianka's speech. In retrospect, this may have been a good thing. It may be that the Lord was providentially protecting me by keeping me from getting tangled up in the controversy. Of course, I do not blame Lamar University for Pianka's comments. My former employer merely happened to be the host for the annual TAS meeting that particular year.

Afterward, I corresponded with Forrest Mims III, the whistleblower of Pianka's statements. Mims attended Pianka's keynote address and was, at the time, the Chairman of the Environmental Science Section of the Texas Academy of Sciences. Though technically an amateur scientist, Mims has stellar credentials.[9] As an interesting aside,

Mims was once denied a permanent position as a columnist at *Scientific American* because of his support for intelligent design.[10] I was able to confirm some of Mims' claims about Pianka. Likewise, a University of Texas–Pan American professor who also attended the keynote address eventually corroborated Mims' statement. He confirmed, as have others, that Mims did *not* exaggerate or take Pianka's statements out of context.[11,12] Furthermore, Pianka has given the same speech many times and has even posted the text online.[13] From his own written words, Pianka clearly thinks most of us dying would be a good thing:

> But notice the estimated population curve with a collapse [of the population]—things are going to get better after the collapse because humans won't be able to decimate the Earth so much. And, I actually think the world will be much better off when only 10 or 20 percent of us are left. It would give wildlife a chance to recover—we won't need conservation biologists anymore. Things are going to get better for the other denizens of Earth as they deteriorate for humans.[14]

Pianka claims that he was not actually advocating that 90% of us die.[15] However, he obviously believes this would be a good thing if it did happen. Also, not too surprisingly Pianka, like many environmentalists, sees carbon dioxide as a form of pollution.[16]

Some are so convinced of global warming that they do not think it is possible for anyone to have good-faith skepticism on the subject. The inflammatory epithet "denier" is often hurled at those who question the consensus view. A number of people, some well known, have proposed that the government punish climate change skeptics.[17,18] One United States Senator proposed using the RICO (Racketeer Influenced and Corrupt Organizations) Act to prosecute climate organizations seen as misinforming the public about climate change.[19]

Oddly enough, some environmentalists argue that changing the planet is bad even if that change results in conditions that most people—even most environmentalists—would consider good. For instance, Eric Pianka thinks that distillation of salt water and an increase in green vegetation are bad things.

> Unlimited cheap clean energy, such as that so ardently hoped for in the concept of cold fusion, would actually be one of the worst things that

> could possibly befall humans. Such energy would enable well meaning but uninformed massive energy consumption and habitat destruction (i.e., mountains would be leveled, massive water canals would be dug, ocean water distilled, water would be pumped and deserts turned into green fields of crops).[20]

Many people, even many environmentalists, assume that greening of the world's deserts would be a good thing. After all, don't government public service announcements continually urge us to "go green"?

Pianka is an evolutionary ecologist in addition to being a herpetologist. Could his evolutionary beliefs be helping to fuel his seemingly extreme views? Such beliefs are common among environmentalists. For instance, the view that affordable, abundant energy is dangerous is widespread. Environmentalist Paul Ehrlich once said, "Giving society cheap, abundant energy would be the equivalent of giving an idiot child a machine gun."[21,22]

In this book, I want to show that the Bible has something to say, either directly or indirectly, about all these concerns: overpopulation, climate change, care of the environment, etc. If one rejects the Bible's testimony, then extreme views like Pianka's are the natural result. This is true even for the seemingly counterintuitive belief that greening of the deserts is a bad thing. Within an evolutionary worldview, such conclusions make perfect sense. But if the evolutionary worldview is wrong, then so are these conclusions. So, I will occasionally refer back to some of these statements, explaining why they are logical within an evolutionary worldview and contrasting them with the biblical worldview.

Hopefully, from the examples I have cited, it should be clear why Christians should be well-informed on this issue. Regardless of what one thinks about global warming, it isn't possible to be indifferent about the issue. As described above, climate alarmists are advocating radical proposals in the name of saving the planet. They are prepared to squelch scientific argument and use the coercive power of the government to ensure

that the "correct" side wins this debate. In their minds, these radical proposals are totally justified because those of us who are skeptical of their claims are endangering the planet and the well-being of living things on it. But if the alarmists are wrong, then the extreme policies they are advocating are unnecessary and have the potential to inflict much needless misery on the human race.

References

1. Schiermeier, Q. 2013. Quandary over Soviet croplands. *Nature.* 504 (7480): 342.
2. Hebert, J. 2014. The Bitter Harvest of Evolutionary Thinking. *Acts & Facts.* 43 (4): 14.
3. Swain, F. Climate change: Could we engineer greener humans? BBC. Posted on bbc.com July 15, 2014, accessed February 20, 2019.
4. Singh, M. Gassy Cows Are Warming the Planet, And They're Here to Stay. Posted on npr.org April 12, 2014, accessed February 20, 2019.
5. Andersen, R. How Engineering the Human Body Could Combat Climate Change. Posted on theatlantic.com March 12, 2012, accessed February 20, 2019.
6. Witt, J. Doctor Doom, Eric Pianka, Receives Standing Ovation from Texas Academy of Science. *Evolution News.* Posted on evolutionnews.com April 3, 2006, accessed May 9, 2018.
7. Pearcey, R. Dr. "Doom" Pianka Speaks. *The Pearcey Report.* Posted on pearceyreport.com April 6, 2006, accessed May 9, 2018.
8. Sandberg, L. Controversial UT professor warns of Earth's end. *The Houston Chronicle.* Posted on chron.com August 27, 2006, accessed May 9, 2018.
9. One can read some of Mims' impressive credentials at his personal website forrestmims.org
10. Mims III, F. The *Scientific American* Affair. Posted on forrestmims.org, accessed June 11, 2019.
11. Helms, H. About Forrest M. Mims III. Posted on forrestmims.org, accessed June 11, 2019.
12. Redford, J. Forrest Mims did not Misrepresent Prof. Eric Pianka's Statements. Web Cite. Posted on webcitation.org April 13, 2006, accessed February 20, 2019.
13. Pianka, E. The Vanishing Book of Life on Earth. School of Biological Sciences at the University of Texas at Austin. Posted on zo.utexas.edu, accessed May 9, 2018.
14. Ibid, 32-33.
15. Pianka, E. The "Controversy" over Eric Pianka's Speech. School of Biological Sciences at the University of Texas at Austin. Posted on zo.utexas.edu, accessed May 9, 2018.
16. Pianka, The Vanishing Book of Life on Earth, 19-20.
17. Chumley, C. K. RFK Jr. wants law to 'punish global warming skeptics.' *The Washington Times.* Posted on washingtontimes.com September 23, 2014, accessed February 20, 2019.
18. Torcello, L. Is misinformation about the climate criminally negligent? *The Conversation.* Posted on theconversation.com March 13, 2014, accessed February 20, 2019.
19. Whitehouse, S. The fossil-fuel industry's campaign to mislead the American people. *The Washington Post.* Posted on washingtonpost.com May 29, 2015, accessed February 20, 2019.
20. Pianka, The Vanishing Book of Life on Earth, 31.
21. Ehrlich was also the author of the 1968 book *The Population Bomb*, which infamously (and incorrectly) predicted mass starvation of humans in the 1970s and 1980s due to overpopulation.
22. Hendrickson, M. A "Mean Green", Obama Wages War Against Cheap Energy. *Forbes.* Posted on forbes.com September 20, 2012, accessed February 20, 2019.

3 Global Warming: A Primer

Summary: Where did climate change alarmism come from? It largely comes from the observation that our planet has gotten warmer over the past 50 years or so. Earth is constantly receiving energy from the sun. Fortunately for us, our atmosphere contains carbon dioxide and other greenhouse agents that trap infrared energy radiated by the earth, preventing much of it from escaping back into space. This trapped energy warms the earth. Without this warming, Earth would be too cold for life. However, some people are concerned that carbon emissions from factories, cars, and other machines are putting too much carbon dioxide into the atmosphere that will trap more energy and increase Earth's temperature to dangerous levels. Ultimately, the climate change debate hinges on something called *climate sensitivity*, the eventual average surface temperature change resulting from changes in carbon dioxide levels. A high climate sensitivity means that increased carbon dioxide will cause subsequent changes that amplify the small amount of initial warming. On the other hand, low climate sensitivity means that the climate responds in such a way as to minimize or reduce any additional warming. Alarmists think climate sensitivity is relatively high, implying an unstable climate, while skeptics think it's relatively low, implying a stable climate. In the following chapters, we'll discuss how Genesis is relevant to this issue.

Data collected at Mauna Loa Observatory (MLO) in Hawaii have played a large part in the global warming controversy. This important scientific research station is located on Hawaii's Big Island (Figure 3.1). Its remote location and high altitude make it an ideal place to make observations of the atmosphere. In fact, I used atmospheric electricity data collected at MLO in my Ph.D. research.[1]

Mauna Loa

Since 1958, MLO scientists have made careful measurements of the amount of carbon dioxide (CO_2) in the atmosphere.[2,3] These measurements (Figure 3.2) show an unmistakable increase in atmospheric CO_2 over the last half century. Many scientists are concerned about this increase. To understand why, we have to discuss some background material.

Figure 3.1. Mauna Loa Observatory in Hawaii

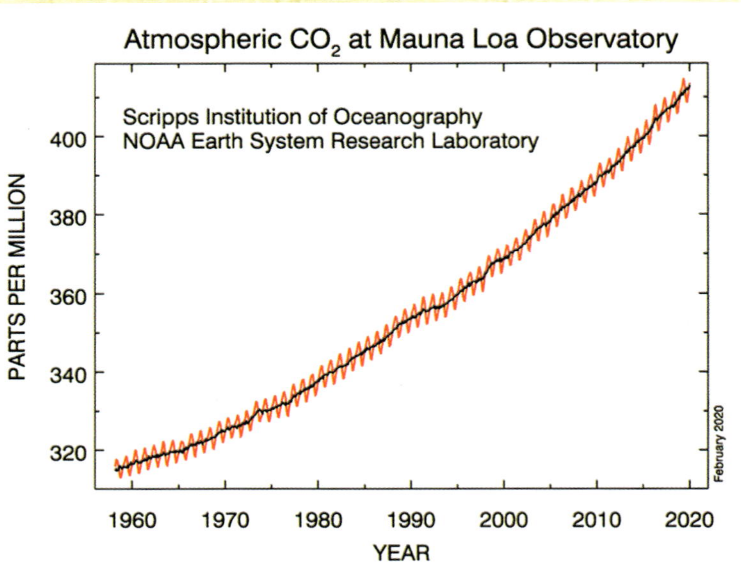

Figure 3.2. Measurements at Mauna Loa Observatory have shown a clear increase in concentrations of atmospheric carbon dioxide.

Electromagnetic Radiation

Visible light, to which our eyes respond, is a form of what physicists call *electromagnetic radiation*. To begin understanding the idea of electromagnetic radiation, we must briefly discuss the concept of a *field*.

The concepts of electric and magnetic fields are somewhat abstract, but most of us have had some experience with them in everyday life, as well as in high school or college physics classes. When you rub a balloon vigorously against a sweater on a cold day, both the sweater and the balloon become charged. The balloon is attracted to the sweater (and vice versa) because the charge on the balloon "senses" the presence of the charge on the sweater. They are able to sense each other because of the electric field caused by the separated charges.

Likewise, if you place a thin glass plate on top of a magnet and then sprinkle iron filings on top of the plate, the filings align themselves along the magnetic field. Just as charged objects can respond to the presence of an electric field, moving charges can respond to the presence of a magnetic field. In fact, there is a kind of symmetry here—

> "Ultimately, the climate change debate hinges on something called *climate sensitivity*, the eventual average surface temperature change resulting from changes in carbon dioxide levels."

charges cause electric fields, and electric fields can "push" charges. Likewise, moving charges cause magnetic fields, and magnetic fields can turn moving charges.

When a charge at rest begins moving, or when a moving charge changes its speed or direction (or both), we say that the charge has *accelerated*. An accelerating charge gives off *electromagnetic radiation*. Electromagnetic radiation is made of oscillating, or vibrating, electric and magnetic fields. These electric and magnetic fields are perpendicular to one another and travel at the speed of light. In a vacuum, this speed is 300,000 kilometers per second. These perpendicular, coupled electric and magnetic fields constitute an *electromagnetic wave*. So, electromagnetic waves radiate outward from accelerating charges.

Most of us are at least somewhat familiar with waves. In a particular material, a wave has a speed that depends on the properties of the material. The speed of the wave is equal to the wavelength (the distance between two adjacent wave crests or troughs) multiplied by the frequency (the number of wavelengths, or wave cycles, that pass a given location per unit time).

The smallest possible unit, or *quantum*, of electromagnetic radiation is the *photon*. The energy of a particular kind of electromagnetic radiation depends on the frequency of the photons of that radiation. A photon with twice the frequency of another photon will have twice as much energy.

Different kinds of electromagnetic radiation travel at the speed of light, but their electric and magnetic fields oscillate at different frequencies (Figure 3.3). Physicists have labeled these different forms of electromagnetic radiation (in order of lowest to highest frequency) as radio waves, microwaves, infrared radiation, visible light, ultraviolet radiation, X-rays, and gamma rays. The colors of visible light are (in order of lowest to highest frequency) red, orange, yellow, green, blue, indigo, and violet. Students often use the memory device

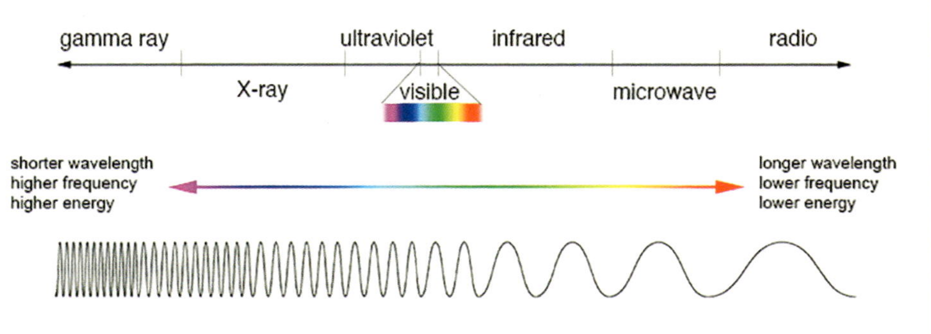

Figure 3.3. All forms of electromagnetic radiation (including visible light) travel at the speed of light, but they have different frequencies of vibration.

ROYGBIV to remember the correct order of the colors. However, visible light only makes up a tiny portion of the *electromagnetic spectrum*.

Ultraviolet (UV) *radiation* has higher frequencies than visible light, which gives it more energy per photon. UV radiation is responsible for giving you a sunburn when you spend too much time in the sun without skin protection. X-rays and gamma rays have even more energy per photon and can cause serious damage to living things. Fortunately for us, God designed Earth's atmosphere to block the most energetic forms of UV radiation, as well as X-rays and gamma rays. Likewise, just beyond the red portion of the visible spectrum is lower-frequency *infrared radiation*. Molecules emit infrared radiation due to their rotational, or vibrational, motions. When you put your hands near a hot stove, you feel the infrared radiation as heat.

Here are three important facts about electromagnetic radiation. First, all objects made of atoms emit electromagnetic radiation. Second, although these objects emit radiation of many different frequencies, the radiation emitted by the object the most (or at the greatest intensity) depends on the object's temperature. For instance, our bodies also emit radiation, but at a body temperature of 98.6°F (37.0°C), this emitted radiation is mainly infrared. Likewise, the earth, with its average surface temperature of 14°C, also emits mainly infrared radiation. You cannot see this infrared radiation with your eyes, but you can see it with infrared goggles. In contrast, the much hotter surface of the sun, at 5,500°C, emits mainly visible light.

Third, the energy radiated by an object in a given amount of time can be mathematically described by its temperature (measured in units called *Kelvins*) raised to the fourth power. This means that if the temperature of an object measured in Kelvins doubles, then the rate at which it radiates electromagnetic radiation will become $2\times2\times2\times2 = 16\times$ greater.

Energy Balance

The earth and its atmosphere (what we call the earth-atmosphere system) is continually receiving electromagnetic energy from the sun. Yet the earth also emits electromagnetic energy. If the amount of energy (in the form of visible and ultraviolet radiation) absorbed by the earth-atmosphere system (per unit time) is equal to the amount of energy (in the form of infrared radiation) emitted (also per unit time), then the earth will be in a state of "energy balance" (Figure 3.4). In other words, energy balance occurs when the rate of incoming energy equals the rate of outgoing energy. When that happens, the earth's temperature will remain constant. What happens if the rate of incoming energy does not balance the rate of outgoing energy?

If the rate of incoming energy is *greater* than the rate of outgoing energy, then this excess of incoming energy will warm the earth, raising its temperature. Remember

Figure 3.4. The earth receives visible and ultraviolet energy from the sun, but it also emits infrared radiation. The amount of energy radiated by the earth per unit time depends on the average surface temperature, which will either rise or fall until a condition of energy balance is achieved: energy received per unit time equals energy emitted per unit time.

"In other words, energy balance occurs when the rate of incoming energy equals the rate of outgoing energy. When that happens, the earth's temperature will remain constant."

that the rate at which energy is radiated goes as temperature raised to the fourth power. This increases the rate at which the earth radiates energy. Theoretically, the rate of outgoing energy will increase until it balances the rate of incoming energy.

Likewise, if the rate of incoming energy is *less* than the rate of outgoing energy, then the earth loses energy. This energy loss causes cooling, which causes the rate of emitted radiation to decrease. In theory, the rate of outgoing radiation will continue to decrease until it matches the lower rate of incoming radiation. So, the earth-climate system is continually trying to achieve energy balance, although continually changing conditions may prevent true energy balance from occurring. But what does this have to do with carbon dioxide and other greenhouse gases?

Greenhouse Gases

Carbon dioxide, like water vapor and clouds, is a "greenhouse agent." This means that carbon dioxide is very good at absorbing and emitting infrared radiation. Most of the energy from the sun is in the form of visible light, infrared radiation, and lower-energy ultraviolet radiation. Half of this energy is in the form of visible light and ultraviolet, and the other half is almost entirely infrared radiation.[4] In fact, the sun emits most of its energy in the form of visible light. Although evolutionists claim that this is because our eyes evolved to make use of this fact, the Lord obviously engineered it this way for our benefit.[5,6]

Because the atmosphere blocks the sun's infrared radiation, nearly all the sun's energy that reaches us is in the form of visible and ultraviolet light. About 30% of this radiant energy goes back into space. It is reflected from clouds or the earth's surface, or it is scattered back into space by molecules in the atmosphere. Of the remaining

70%, the atmosphere absorbs 20% and the earth's surface absorbs 50%.[6]

The earth's surface absorbs the energy of this visible light and ultraviolet radiation. This makes the earth's surface warmer. Then the earth's surface radiates energy, most of which is in the form of infrared radiation.

The molecules of greenhouse agents (including carbon dioxide) absorb and emit infrared radiation very efficiently. Hence, they absorb the infrared radiation coming up from the earth's surface. Then they radiate some of this infrared radiation upward and some of it back down to the earth's surface. The outgoing infrared radiation is lost to space, but this downward infrared radiation warms the earth and the lower atmosphere.

Fortunately, our atmosphere contains greenhouse agents. If Earth had no atmosphere, or an atmosphere with molecules that only weakly interacted with infrared radiation, then the earth would radiate all, or nearly all, its emitted energy back into space, causing the earth's surface to be much colder.[7] The earth's average surface temperature is 14°C (57°F). However, basic physics calculations show that without a greenhouse gas-containing atmosphere, the temperature would be a much colder -19°C (-3°F)![8-10]

The Controversy

To review, greenhouse agents warm the earth. But since the amount of atmospheric carbon dioxide (a greenhouse gas) is increasing, then could things get too hot?

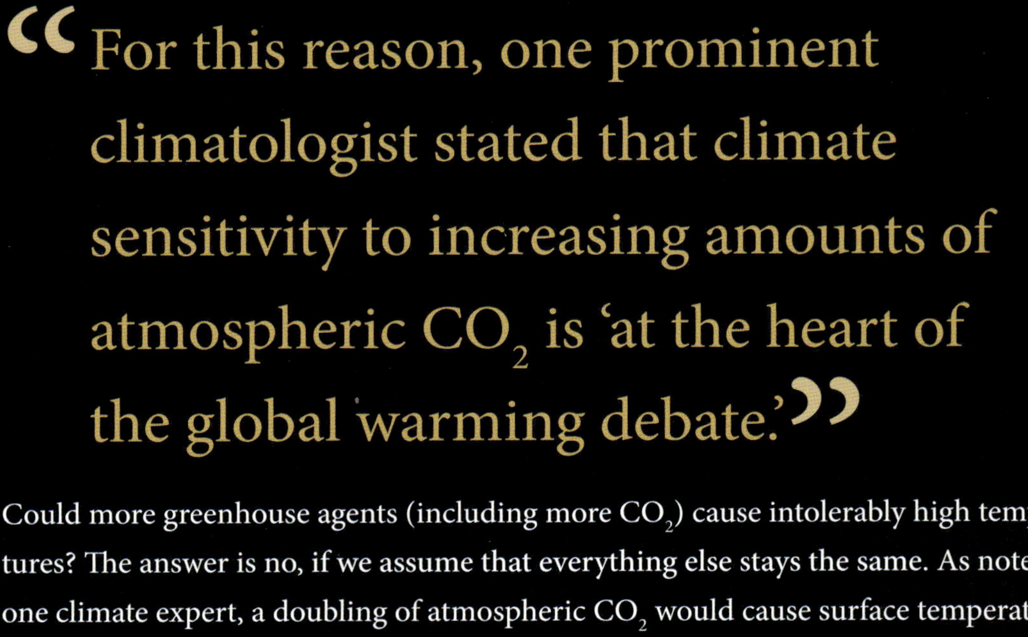

> “For this reason, one prominent climatologist stated that climate sensitivity to increasing amounts of atmospheric CO_2 is ‘at the heart of the global warming debate.’”

Could more greenhouse agents (including more CO_2) cause intolerably high temperatures? The answer is no, if we assume that everything else stays the same. As noted by one climate expert, a doubling of atmospheric CO_2 would cause surface temperatures to increase by less than one degree Celsius (1.8°F).[11]

However, if the amount of atmospheric CO_2 doubles, then everything else probably won’t stay the same. The earth’s atmosphere responds in complicated ways to this increased warming. Some of these responses, called *positive feedbacks*, could enhance any warming caused by increased CO_2, and some of these responses, called *negative feedbacks*, could minimize additional warming.

The real issue in the climate change or global warming debate is this:

How sensitive is the earth's climate to changes in CO_2? How sensitive is it to other changes? Scientists who are more concerned about global warming think the earth's climate sensitivity is high, while those who are less concerned think it is low. For this reason, one prominent climatologist stated that climate sensitivity to increasing amounts of atmospheric CO_2 is "at the heart of the global warming debate."[12]

Now that we have discussed the basics of the global warming issue, we will discuss why the book of Genesis is so important for correctly weighing the evidence on this issue.

References

1. Hebert, J. 2011. Atmospheric Electricity Data from Mauna Loa Observatory: Additional Support for a Global Electric Circuit-Weather Connection? Ph.D. dissertation. University of Texas at Dallas.
2. Monroe, R. The History of the Keeling Curve. Scripps Institution of Oceanography. Posted on scripps.ucsd.edu April 3, 2013, accessed February 20, 2019.
3. Interestingly, Forrest Mims III, mentioned in chapter 2, once worked at Mauna Loa Observatory.
4. Lutgens, F. K. and E. J. Tarbuck. 2010. *The Atmosphere: An Introduction to Meteorology*, 11th ed. New York: Prentice Hall, 47.
5. Hewitt, P. G. 2006. *Conceptual Physics*, 10th ed. San Francisco: Pearson/Addison Wesley, 518.
6. Lugens and Tarbuck, *The Atmosphere*, 56.
7. Ibid, 54.
8. Global Warming. NASA Earth Observatory. Posted on earthobservatory.nasa.gov, accessed February 20, 2019.
9. In fact, if the earth didn't lose heat due to a phenomenon called convection, the warming due to the greenhouse effect would be even greater than it is. See Lindzen, R. S. 1990. Some Coolness Concerning Global Warming. *Bulletin of the American Meteorological Society.* 71 (3): 288-299.
10. Calculating Planetary Energy Balance and Temperature. UCAR Center for Science Education. Posted on scied.ucar.edu, accessed February 20, 2019.
11. Spencer, R. Global Warming 101. Posted on drroyspencer.com, accessed February 20, 2019.
12. Curry, J. Lewis and Curry: Climate sensitivity uncertainty. Posted on judithcurry.com September 24, 2014, accessed January 4, 2017.

4 The Genesis Flood in Earth History

Summary: As we will see later, different responses to possible climate change can be traced back to different assumptions about Earth's past. When secular scientists try to explain huge changes in Earth's climate in the past—such as Ice Ages—they use a model that is too weak, and so they must find something else to blame. However, the global Flood of Genesis provides the starting point to a much stronger model of explaining Earth's climate. Evidence for the global Flood includes the Bible's testimony and the fact that sedimentary (water-deposited) rocks cover 75% of Earth's land surface, thin continent-wide sedimentary deposits were catastrophically deposited over vast regions at the same time, and fossils of marine and land creatures are often found buried together, as well as the large number of Flood accounts and legends found throughout the world, that are remarkably similar to the account in Genesis. The Flood provides a plausible trigger for the Ice Age, which was a huge change in Earth's climate. Since a catastrophic, never-to-be-repeated event (the Flood) was required to bring about such drastic climate change (the Ice Age), we do not need to worry that noncatastrophic causes such as relatively slow increases in atmospheric carbon dioxide will result in a future climate catastrophe.

To understand why Genesis is so relevant to the issue of climate change, it is necessary to step back and look at the big picture of Earth history. Earth scientists can be grouped into two big categories: creation scientists and uniformitarians. Creation scientists accept the Bible as the inerrant Word of God. For this reason, creation scientists accept the Bible's claim that God created the universe in six days within the recent past, about 6,000 years ago. We also accept the Bible's claim that God used a cataclys-

Figure 4.1. The Genesis Flood was a true global flood, not merely a local flood as some claim

mic Flood to destroy a sinful world in the days of Noah (Figure 4.1).

The Genesis Flood makes good sense of the rocks and fossils, as well as the earth's overall geography. Today, 71% of the earth's surface is still underwater, with very little land visible in the central Pacific Ocean (Figure 4.2). If the earth's topography were smoothed out, this water is sufficient to cover everything to a depth of more than one and a half miles.[1] Creation scientists think the topography of the pre-Flood world was less pronounced than today's topography. This means that the oceans would have been shallower and the mountains not as tall as today's mountains. At the end of the Flood, mountain uplift and deepening of the ocean basins allowed the floodwaters to drain off the continents.

"Earth scientists can be grouped into two big categories: creation scientists and uniformitarians."

Likewise, sedimentary rocks cover the continents. Sedimentary rocks are the lithified or hardened remnants of water-deposited sediments. Although thin compared to the radius of the earth, these rock layers can be thousands of feet thick (Figure 4.3).[2] Within these water-deposited rocks are the fossilized remains of billions of plants and animals. These remains often

Figure 4.2. More than 70% of Earth's surface is submerged, and very little land at all is visible in the central Pacific Ocean

show evidence of rapid and catastrophic burial.

In addition, people groups throughout the world have hundreds of stories or legends of a great flood that annihilated the human race except for one man and his family.[3-5] These recollections are very similar to the account of the Flood described in Genesis 6–9.

However, the second group of scientists holds to a philosophy called *uniformitarianism*, summarized by the motto "The present is the key to the past." Uniformitarian scientists claim that natural processes, operating at more or less the rates and intensities they do today, are capable of explaining both our existence and Earth history apart from the intervention of a supernatural Creator. Uniformitarians reject out-of-hand the possibility that the Bible's account of history could be true. For them, serious consideration of a supernatural creation and a literal worldwide flood is completely out of the question.

Figure 4.3. Creation scientists argue that sedimentary strata, such as those showcased at Grand Canyon, were deposited during the Genesis Flood.

For the creationist, this rejection of biblical history is not surprising. In fact, the apostle Peter long ago prophesied such scoffers would come in the last days:

> Knowing this first: that scoffers will come in the last days, walking according to their own lusts, and saying, "Where is the promise of His coming? For since the fathers fell asleep, all things continue as they were from the beginning of creation." For this they willfully forget: that by the word of God the heavens were of old, and the earth standing out of water and in the water, by which the world that then existed perished, being flooded with water. (2 Peter 3:3-6)

Peter says these scoffers will deny three important truths. First, they will deny that God created the universe (v. 5). Second, they will deny that God destroyed the world with the Flood in the days of Noah (v. 6). Third, they will scoff at the promised return of Jesus Christ (v. 4) to both judge and punish the ungodly (v. 7) and ultimately establish "new heavens and a new earth in which righteousness dwells" (v. 13).[6]

Before we can draw conclusions about possible future climate change, we need to correctly understand *past* climate change. Therefore, it is important to see which point of view is more credible. Are the rocks and fossils a record of millions of years of evolutionary change, or are they the result of the Genesis Flood?

Sedimentary Rocks

Because the Genesis Flood is so critical to correctly understanding Earth history, we'll briefly describe how the Genesis Flood does a better job explaining the rocks and fossils than evolutionary assumptions do.

Sedimentary rocks cover about 75% of Earth's land surface.[7] ICR geologist Dr. Tim Clarey told me that geologists generally agree that 90 to 95% of sedimentary rocks are water-deposited, and this agrees with a figure given to me by a secular geologist in a private conversation. In general, sedimentary rocks are classified as sandstones (20 to 25%), mudrocks (65%), or carbonate rocks (10 to 15%).[8] Sandstones are composed of cemented sand particles, and mudrocks are composed of clays or mud. Carbonate rocks are chemical precipitates and the remains of marine organisms such as foraminifera and corals.

Uniformitarians agree that most sedimentary rocks formed from water-deposited sediments, but they strenuously object to the claim that these rocks are the result of the Flood. Instead, they have long claimed that mudrocks formed in still, placid lake environments. Obviously, these are not the environments one would associate with a catastrophic global deluge. In addition, mudcracks within mudstones were seen as evidence that the mud was exposed to the air for a sufficient time to dry out—again, not something that one would associate with a global flood.

However, recent experiments showed that mudrocks can form under fast-moving water currents.[9,10] Further-

> "The large heights of cross-beds in Grand Canyon sandstones imply fast deep-water speeds of three to five feet per second."

more, some features within mudstones are inconsistent with slow and gradual deposition in a quiet environment.[11] Also, mudcracks can form in already-buried rocks.[12] Therefore, it is a mistake for uniformitarians to assume that mudrocks are incompatible with the Flood.

Uniformitarians have also argued that sandstone rocks, such as Grand Canyon's Coconino Sandstone (Figure 4.4), were formed in air by migrating sand dunes. If true, this would be hard to reconcile with a global flood. However, sandstones can form in a watery environment. Many footprint trackways are preserved in sandstone (Figure 4.5). Creation scientists think submerged or partially submerged animals made these tracks as they moved upstream in an attempt to escape the waters.[13] Likewise, the relatively shallow inclination (25° or less, measured from the horizontal) of cross-beds within the Coconino are more consistent with underwater deposition (Figure 4.6).[14,15]

In spite of this evidence for watery deposition, most uniformitarian geologists reject this interpretation. The large heights of cross-beds in Grand Canyon sandstones imply fast deep-water

Figure 4.4. Cross-beds within Grand Canyon's Coconino Sandstone

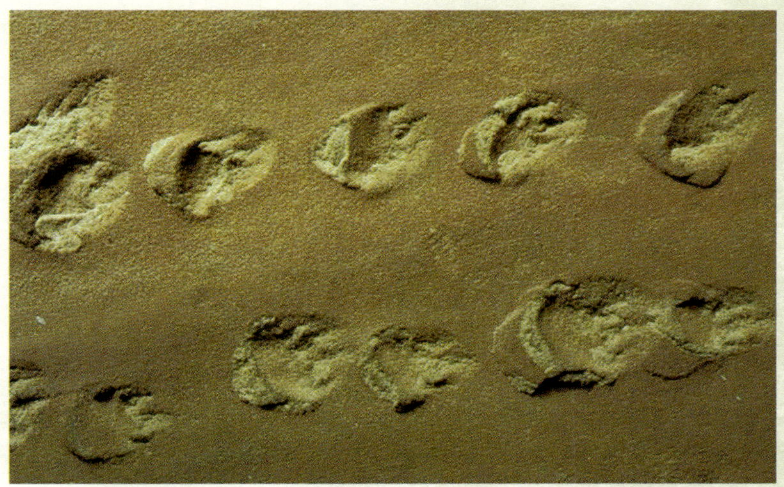

Figure 4.5. Animal trackway preserved in an inclined cross-bed from the Coconino Sandstone

Figure 4.6. Cross-bed inclinations of 25° or less are generally indicative of underwater deposition

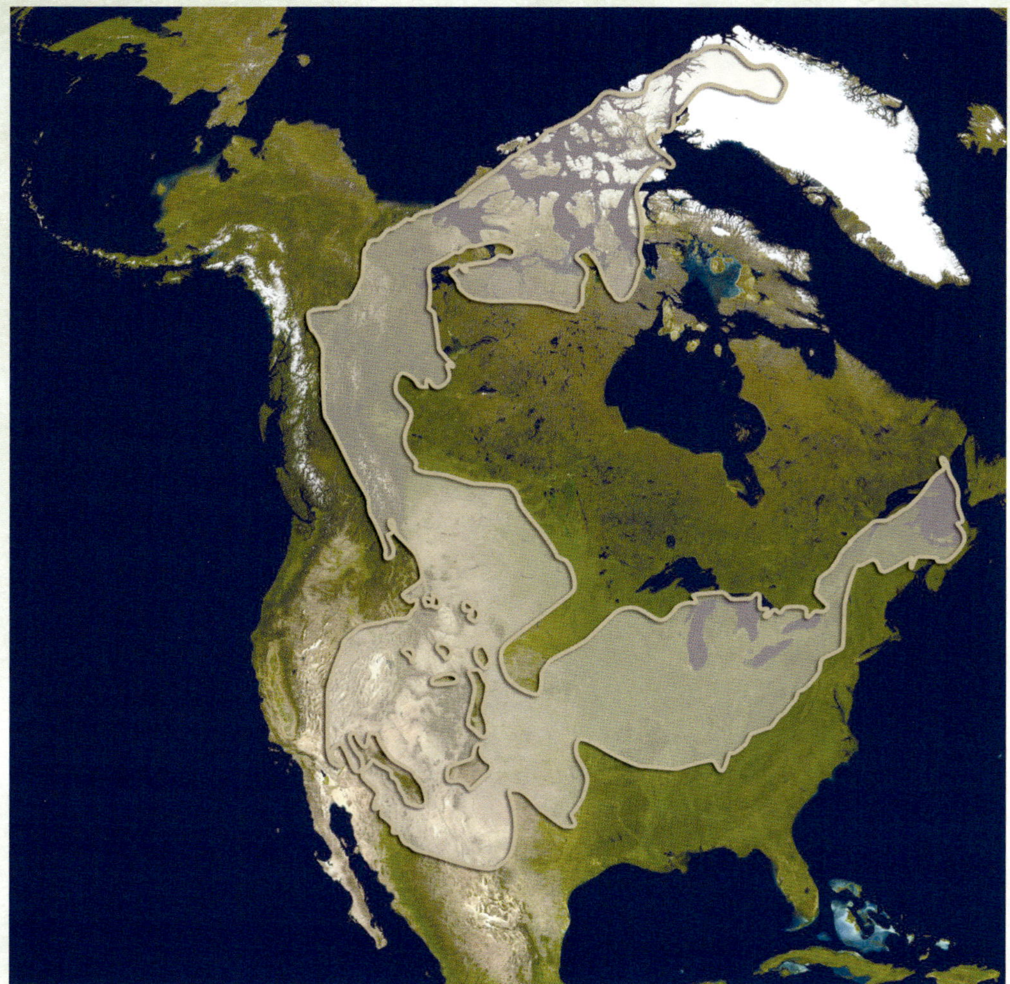

Figure 4.7. The Tapeats Sandstone is a continuous sedimentary sequence that covers most of North America

speeds of three to five feet per second. Since these deep-water speeds rarely, if ever, occur today, uniformitarians (who maintain that the present is the key to the past) have a hard time believing the cross-beds were formed by water currents. Creation scientists point out that high-speed deep-water currents could be expected to occur during the Flood. In fact, most of the sandstones in the Grand Canyon probably formed this way, including the Tapeats Sandstone, the Supai Group sandstones, the Toroweap sandstones, and the sandstones within the Kaibab Limestone.[16]

Moreover, these sedimentary rock units are enormous, often cover large sections of the continents, and even stretch from continent to continent. For instance, although the Tapeats Sandstone is one of the rock units within the Grand Canyon, this particular sandstone (Figure 4.7) covers a large fraction of the continental United States, much of western Canada, and even extends into Greenland.[17] Uniformitarians cannot reasonably explain the enormous sizes of these rock units.

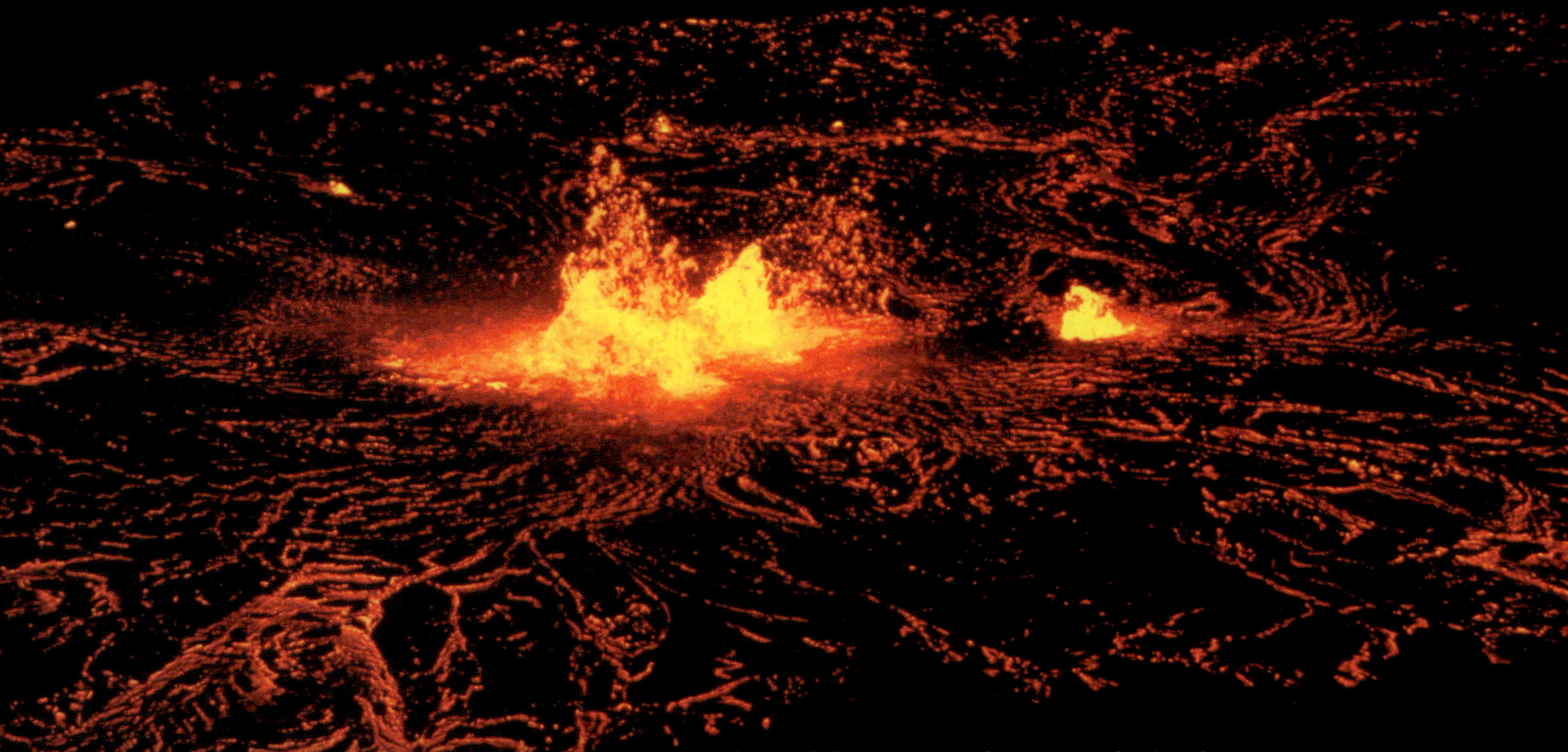

In a similar way, secular geologists long claimed that most carbonate rocks (such as limestone) formed slowly in quiet, low-energy environments. Experiments have shown this claim to be incorrect.[18] Also, it is intriguing that carbonate rocks contain the shells of marine organisms, yet many of these rocks are found on land. Of course, this is to be expected during a global flood.

For these reasons, creation scientists argue that these sedimentary rocks are the result of the Genesis Flood. Of course, this was the standard geological interpretation until the rise of uniformitarianism in the 18th and 19th centuries.

Another Objection to a Global Flood

However, a uniformitarian might object that if the Genesis Flood really did cover the entire surface of the planet (Genesis 7:19-24), then why is only 75% of the earth's land surface covered by sedimentary rocks? Why is this coverage not 100%?

That question is easy to answer. In some cases, especially on high ground, the sediments deposited by the floodwaters were quite thin. The receding floodwaters could

Figure 4.8. A volcanic eruption in Hawaii. Volcanism during the Genesis Flood dwarfed any volcanic activity occurring today.

easily have eroded away these thin sedimentary layers. In fact, the receding floodwaters were capable of eroding away thick sedimentary deposits as well. In some cases, erosion removed thousands of feet of sediment.[19,20] As we shall see later, geological clues within the rocks are consistent with rapid, catastrophic erosion such as one would expect during the latter stages of the Flood. This erosion left original pre-Flood basement rock exposed on the continents.

> "It is common for marine and land fossils to be found together."

Also, volcanic rocks formed during the Flood, not just sedimentary rocks. The scale of volcanic activity during the Flood dwarfs anything that we observe today (Figure 4.8). Over a dozen extremely large volcanic rock units known as *large igneous provinces* (LIPs) exist throughout the world. To put the sizes of these volcanic eruptions into perspective, uniformitarians often associate LIPs with mass extinction events. Volcanism on such a scale does not occur in the present, but it could be expected to have occurred during the Genesis Flood, when "all the fountains of the great deep were broken up" (Genesis 7:11).

Therefore, it is not surprising that Earth's land surface is not completely covered by sedimentary rocks. Erosion by receding floodwaters removed the uppermost sediments, leaving underlying rocks exposed on the continents. In most cases, this exposed rock was original pre-Flood basement rock. However, in some cases this exposed rock was volcanic rock formed during the Flood cataclysm.

Figure 4.9. The Genesis Flood provided the conditions necessary to quickly bury and fossilize trillions of plants and animals

Dinosaur Fossils and the Flood

Of course, trillions of fossils are entombed within these sedimentary rock units (Figure 4.9). These fossils show evidence of violent death and rapid burial. In fact, fossilization requires rapid burial to protect an organism's remains from decay or scavenging by predators. It is common for marine and land fossils to be found together. In fact, the vast majority of the fossils are marine invertebrates, such as clams.[21] Yet these marine vertebrate fossils are found on land. Why? The Genesis Flood provides the obvious answer.

Dinosaur fossils give us an idea of the extent and violence of the Genesis Flood. The fossils of dinosaurs (land animals) are often found with the fossils of marine creatures.[22] For instance, a Canadian oil company worker found an ankylosaur dinosaur fossil within marine sediments usually associated with clams and ammonites.[23] This common mixing of terrestrial and aquatic fossils is more consistent with a rapid, catastrophic flood than slow, gradual "marine transgressions" supposedly occurring over millions of years. Likewise, paleontologists often find dinosaur fossils in large bone beds. Even uniformitarians acknowledge that the dinosaurs within at least some of these bone beds died by drowning, although they prefer to attribute this to a local, rather than a global, flood.[24-26]

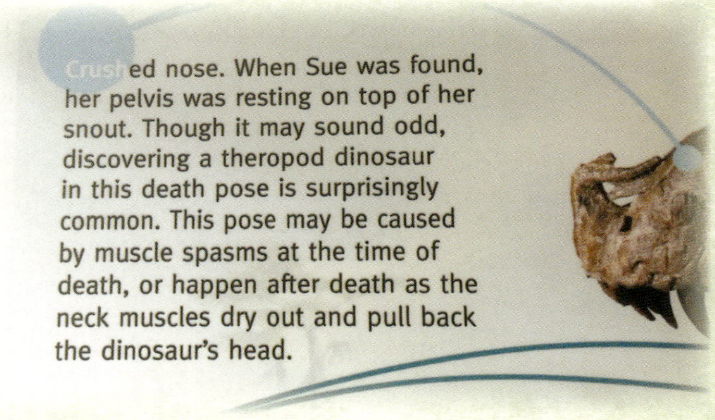

Dinosaur fossils are often incomplete, consistent with violent transport. In fact, sauropod dinosaur fossils are often found without the heads attached.[27] According to the placard at the Chicago Field Museum (Figure 4.10), the famous *T. rex* fossil Sue was found with "her pelvis resting on top of her snout." What kind of conditions fold a giant, seven-ton *T. rex* in half?

Figure 4.10. Placard for the *T. rex* fossil Sue at the Chicago Field Museum. Is it really reasonable to think that mere muscle spasms folded this giant creature in half?

These factors are consistent with rapid burial in a watery catastrophe. Of course, if these dinosaurs really did perish during the Genesis Flood, then it follows that there was never an "age of dinosaurs" when dinosaurs ruled the earth. Dinosaurs, along with all the other land animals, were created with Adam and Eve on Day 6 of the creation

Figure 4.11. Mesopotamian cylinder seal with depiction of dragons

Figure 4.12. Apparent depiction of a *Stegosaurus* at Ta Prohm temple, Siem Reap, Cambodia

Figure 4.13. Original soft tissue preserved in dinosaur and other "prehistoric" fossils is very strong evidence that these fossils are much younger than claimed by secular scientists

week (Genesis 1:24-31). The dinosaurs not taken by Noah on board the Ark would have perished during the Flood, and their fossilized remains resulted from rapid burial within Flood sediments.

Interestingly, many cultures have recollections of dragons, which often sound very similar to the creatures we now call dinosaurs, as well as to other extinct flying and swimming reptiles such as pterosaurs and mosasaurs. Ancient historians, including the writers of Scripture, often described these dragons very matter-of-factly.[28,29] Ancient peoples even made images of these "dragons" that are often accurate depictions of dinosaurs. Figures 4.11 and 4.12 are two such depictions. Even evolutionists do not dispute the authenticity of these pieces of artwork.

Likewise, the preservation of original biomolecules within dinosaur bones is a powerful argument that these creatures lived fairly recently (Figure 4.13). Extrapolations from lab experiments indicate that such biological materials should not survive even a million years under reasonable temperatures.[30-32]

An Evolutionary Fossil Sequence?

One common objection to a Flood interpretation of the rocks and fossils is the supposed order of those fossils in the rocks. Uniformitarian geologists would have us believe there is an orderly, simple-to-complex sequence in the fossils, with the oldest and "simplest" creatures in the deepest sedimentary rocks and the most recently evolved creatures in the uppermost rock layers.

Creation scientists do not dispute that there is a general trend in the fossils, but we would argue that the trend is the result of the floodwaters encroaching upon different habitats and environments. Hence, one would expect to see shallow marine organisms in the deepest sediments, then creatures from coastal habitats, then creatures from higher elevations, etc. Such a general trend does exist, but it contains many surprises for uniformitarians and evolutionists. For instance, a 1992 article in the official publication of the Smithsonian Institution describes what appear to be bird, bear, and ape tracks in Permian rock (supposedly 270 million years old) in New Mexico. Doug Seward reported:

> The fossil tracks that [Jerry] MacDonald has collected include a number of what paleontologists like to call problematica. On one trackway, for example, a three-toed creature apparently took a few steps, then disappeared—as though it took off and flew. "We don't know of any three-toed animals in the Permian," MacDonald points out. "And there aren't supposed to be any birds." He's got several tracks where creatures appear to be walking on their hind legs, others that look almost simian. On one pair of siltstone tablets, I notice some unusually large, deep and scary-looking footprints, each with five arched toe marks, like nails. I

> comment that they look just like bear tracks. "Yeah," MacDonald says reluctantly, "they sure do." Mammals evolved long after the Permian period, scientists agree, yet these tracks are clearly Permian.[33]

By uniformitarian reckoning, both the Permian rocks and the creatures that made these tracks are older than the dinosaurs. Of course, this is a major evolutionary contradiction—birds, apes, and bears supposedly evolved long *after* the dinosaurs went extinct! It is interesting that paleontologists even have a name for such anomalies: *problematica*. This suggests that such out-of-place (or should we say out-of-time?) fossils may be more common than most people think.

Another such anomaly involved bird-like tracks in late Triassic rocks described in a 2002 article in *Nature*.[34] The article's authors admitted that the tracks had clear bird-like characteristics. However, they could not bring themselves to admit that they were bird tracks since birds supposedly did not appear until another 55 million years later in the Jurassic. However, the authors retracted the article in 2013, claiming that the rocks containing the bird-like tracks were really from the late Eocene.[35] This made the tracks 180 million years younger than originally believed, putting them long after the supposed Jurassic appearance of birds. With this awkward out-of-sequence "problem" solved, the authors had no problem admitting that the tracks really *had* been made by birds.[36]

"It is interesting that paleontologists even have a name for such anomalies: *problematica*. This suggests that such out-of-place (or should we say out-of-time?) fossils may be more common than most people think."

Likewise, paleontologists have repeatedly found dinosaur remains in rocks dated younger than 66 million years old despite textbook orthodoxy that the dinosaurs were wiped out by a meteorite impact 66 million years ago.[37,38] Not surprisingly, such claims have been quite controversial, and secular scientists usually claim that these out-of-place dinosaur fossils were somehow "reworked" into younger sediments.[39]

Similarly, new fossil discoveries have forced evolutionists to retract claims about the first appearances of certain organisms. Evolutionists once claimed grasses did not appear until after the dinosaurs became extinct, but this is now known to be false.[40]

Also, by evolutionary reckoning it is only after the extinction of the dinosaurs 66 million years ago that mammals became widespread. But paleontologists have found many, perhaps hundreds, of mammal species in the rock layers containing dinosaurs. Yet museum exhibits or dioramas almost never display these mammal species.[41,42] So although the fossil sequence may be more orderly than evolutionists claim it should be if the Flood had happened, it is certainly *less* orderly than they expect. Indeed, given the above examples some of the apparent order of the fossil record is likely the result of circular reasoning, since problematica such as these are either ignored or explained away.[43, 44]

Of course, there is another very big problem with an evolutionary interpretation of the fossils. The transitional forms that one would expect if evolution were true are

conspicuously absent. Although evolutionists claim transitional forms do exist, the examples they cite are uncertain at best and disputed even by other evolutionists.[45]

Hence, the rocks and fossils were the result of the Genesis Flood, not slow and gradual processes over millions of years. As we shall see, the reality of the Genesis Flood has profound implications for understanding both past and future climate change.

References

1. Walker, T. 2008. Where did all the water go? *Creation.* 30 (3): 41.
2. The average thickness of sedimentary rocks on the continents is about 1.8 kilometers. See Beck, K. C. et al. Sedimentary rock. *Encyclopaedia Britannica.* Posted on britannica.com, accessed November 2, 2016.
3. Morris, J. D. 2001. Why Does Nearly Every Culture Have a Tradition of a Global Flood? *Acts & Facts.* 30 (9).
4. Morris, J. D. 2014. Traditions of a Global Flood. *Acts & Facts.* 43 (11): 15.
5. Cooper, B. 2011. *The Authenticity of the Book of Genesis.* Portsmouth, Hampshire, UK: Creation Science Movement.
6. Although there were undoubtedly also such scoffers in Peter's time, his prophecy seems to be especially relevant for today.
7. Thompson, G. R. and J. Turk. 1997. *Modern Physical Geology*, 2nd ed. Fort Worth, TX: Saunders College Publishing, 132.
8. Nelson, S. A. Geology 212 Petrology Notes. Tulane University. Posted on tulane.edu, accessed November 15, 2016.
9. Schieber, J., J. Southard, and K. Thaisen. 2007. Accretion of Mudstone Beds from Migrating Floccule Ripples. *Science.* 318 (5857): 1760-1763.
10. Walker, T. Mud experiments overturn long-held geological beliefs: A call for a radical reappraisal of all previous interpretations of mudstone deposits. Creation Ministries International. Posted on creation.com January 9, 2008, accessed November 15, 2016.
11. As waters clear, scientists seek to end a muddy debate. *Phys.org.* Posted on phys.org December 13, 2007, accessed November 15, 2016.
12. Hoesch, W. A. 2006. Mudcracks and the Flood. *Acts & Facts.* 35 (11).
13. Austin, S. A. 1995. *Grand Canyon: Monument to Catastrophe.* Santee, CA: Institute for Creation Research, 28-32.
14. Ibid, 32.
15. Thomas, B. 2014. Do Sand-Dune Sandstones Disprove Noah's Flood? *Acts & Facts.* 43 (9): 18-19.

16. Austin, *Grand Canyon: Monument to Catastrophe*, 33-35.
17. Morris, J. D. 2012. *The Global Flood*. Dallas, TX: Institute for Creation Research, 149.
18. Clarey, T. 2018. Rapid Limestone Deposits Match Flood Account. *Acts & Facts*. 47 (6): 9.
19. Oard, M. J. 2011. The remarkable African Planation Surface. *Journal of Creation*. 25 (1): 111-122.
20. Baumgardner, J. 2005. Recent Rapid Uplift of Today's Mountains. *Acts & Facts*. 34 (3).
21. Morris, J. D. and F. J. Sherwin. 2010. *The Fossil Record: Unearthing Nature's History of Life*. Dallas, TX: Institute for Creation Research, 41.
22. Clarey, T. 2015. Dinosaurs in Marine Sediments: A Worldwide Phenomenon. *Acts & Facts*. 44 (6): 16.
23. Gordon, J. Rare dinosaur found in Canada's oil sands. Reuters. Posted on ca.reuters.com March 25, 2011, accessed November 15, 2016.
24. Ryan, M. J. et al. 2001. The Taphonomy of a *Centrosaurus* (Ornithischia: Ceratopsidae) Bone Bed from the Dinosaur Park Formation (Upper Campanian), Alberta, Canada, with Comments on Cranial Ontogeny. *Palaios*. 16 (5): 482-506.
25. Chiba, K. et al. 2015. Taphonomy of a Monodominant *Centrosaurus apertus* (Dinosauria: Ceratopsia) Bonebed from the Upper Oldman Formation of Southeastern Alberta. *Palaios*. 30 (9): 655-667.
26. Thomas, B. Canadian 'Mega' Dinosaur Bonebed Formed by Watery Catastrophe. *Creation Science Update*. Posted on ICR.org July 13, 2010, accessed November 15, 2016.
27. Black, R. The Mystery of the Missing Brontosaurus Head. *Smithsonian Magazine*. Posted on smithsonianmag.com August 23, 2010, accessed November 15, 2016.
28. For example, see the King James Version translations of Job 40:15-24; Job 41; Psalm 74:14; Psalm 91:13; Psalm 104:26; Isaiah 14:29; Isaiah 27:1; Isaiah 30:6; Isaiah 51:9; Jeremiah 51:34; and Ezekiel 29:3. Although newer Bible translations will sometimes translate the word "dragon" as "jackal," this is not justified because some of these passages are clearly describing reptilian and/or monstrous creatures. Bible translators of the past had no problem acknowledging the past existence of dragons.
29. Cooper, B. 1995. *After the Flood: The Early Post-Flood History of Europe Traced Back to Noah*. Chichester, West Sussex, UK: New Wine Press, 130-161.
30. Thomas, B. Dinosaur Soft Tissue Finally Makes News. *Creation Science Update*. Posted on ICR.org December 2, 2009, accessed December 3, 2018.
31. Thomas, B. Dinosaur Soft Tissue Issue Is Here to Stay. *Creation Science Update*. Posted on ICR.org September 1, 2009, accessed December 3, 2018.
32. Thomas, B. Scientists Broom Challenging Discoveries Beneath 'Contamination' Rug. *Creation Science Update*. Posted on ICR.org May 15, 2013, accessed December 3, 2018.
33. Stewart, D. 1992. Petrified Footprints: A Puzzling Parade of Permian Beasts. *Smithsonian Magazine*. 23 (4): 70-79.
34. Melchor, R. N., S. de Valais, and J. F. Genise. 2002. Bird-like fossil footprints from the Late Triassic. *Nature*. 417: 936-938.
35. Melchor, R. N., S. de Valaia, and J. F. Genise. 2013. Retraction: Bird-like fossil footprints from the Late Triassic. *Nature*. 501: 262.
36. Melchor, R. N., R. Buchwaldt, and S. Bowring. 2013. A Late Eocene date for Late Triassic bird tracks. *Nature*. 495 (7441): E1-E2.
37. Rigby, J. K., Jr. et al. 1987. Dinosaurs from the Paleocene Part of the Hell Creek Formation, McCone County, Montana. *Palaios*. 2 (3): 296-302.
38. Fassett, J. E. 2008. New geochronologic and stratigraphic evidence confirms the paleocene age of the dinosaur-bearing ojo alamo sandstone and animas formation in the San Juan Basis, New Mexico and Colorado. *Palaeontologia Electronica*. 12 (1): 3A.
39. Buck, B. J. et al. 2004. "Tertiary Dinosaurs" in the Nanxiong Basin, Southern China, Are Reworked from the Cretaceous. *Journal of Geology*. 112: 111-118.
40. Neergaard, L. Dinosaur poop shows grass is older than it seems. *Seattle Pi*. Posted on seattlepi.com November 17, 2005, accessed November 15, 2016.
41. Werner, C. 2008. *Evolution: The Grand Experiment, Vol. 2—Living Fossils*. Green Forest, AR: New Leaf Press.
42. Batten, D. 2011. Living Fossils: a powerful argument for creation. *Creation*. 33 (2): 22-23.
43. Oard, M. J. 2011. Taxonomic manipulations likely common. *Journal of Creation*. 25 (3): 15-17.
44. Oard, M. J. 2011. *Dinosaur Challenges and Mysteries: How the Genesis Flood Makes Sense of Dinosaur Evidence—Including Tracks, Nests, Eggs, and Scavenged Bonebeds*. Atlanta, GA: Creation Book Publishers, 156-162.
45. Morris and Sherwin, *The Fossil Record*, 129-177.

5 Plate Tectonics and the Flood

Summary: The Genesis Flood offers a good explanation for plate tectonic movement. Earth's crust is separated into seven or eight large plates. The continents are made of granite, and the ocean floors are made of basalt. The creationist theory of catastrophic plate tectonics makes better sense of geophysical data than the standard slow and gradual plate tectonics theory.

Before the Flood, all of the continents were joined together. During the Flood, the continents rapidly moved apart. Plates were rapidly subducted (pulled into Earth's mantle), destroying the pre-Flood ocean floor. Magma seeped up between the separating plates at the mid-ocean ridges, rapidly forming new ocean floor. This hot, new seafloor had lower density than the surrounding rock and buoyed upward, pushing enormous amounts of water onto the continents. The heat from this rapidly formed seafloor greatly warmed the world's oceans, providing one of the key ingredients needed for an Ice Age. In the next chapter, we'll look at what Earth's climate might have been like before the Flood.

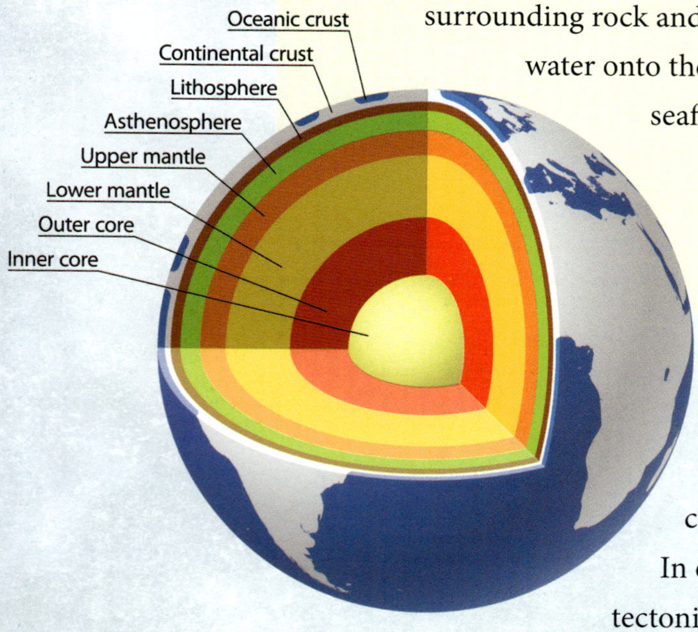

Oceanic crust
Continental crust
Lithosphere
Asthenosphere
Upper mantle
Lower mantle
Outer core
Inner core

Most creation scientists are convinced that instead of multiple continents, a single supercontinent existed in the pre-Flood world. This supercontinent broke apart during the Flood cataclysm. In other words, most creation scientists think plate tectonics played a major role in the Genesis Flood. People not well versed in creation literature would probably assume that plate tectonics is a problem for Flood geology. After all, our Earth science and geology teachers taught us that the continents have slowly drifted apart over millions of years. In many peoples' minds, plate tectonics, like the supposed age of dinosaurs, is synonymous with millions of years of Earth history. Likewise, some might think that past flips or reversals of the earth's magnetic field, during which the earth's north and south magnetic poles switched, are also problematic for creation scientists since secular scientists claim these reversals occurred over many millions of years.

However, this is not the case. A particular Flood model called *catastrophic plate tectonics* (CPT) makes good sense of both the evidence for plate tectonics and magnetic reversals. Of course, the ability to explain a wide array of seemingly unrelated phenomena with a minimum of additional hypotheses is a characteristic of a good scientific model.

Plate Tectonics

Earth's interior consists of three layers. The outermost layer is a thin crust. Below the crust is a thick mantle made of silicate rock. Although temperatures in the mantle are very high, high pressures in the mantle usually keep the mantle rock from melting, causing the mantle to behave as a solid that slowly deforms over time. Below the mantle is a small core thought to be composed mainly of iron and nickel. The core consists of both an outer liquid layer and a solid inner layer.

Earth's crust and upper mantle together constitute what geophysicists call the *lithosphere*. The lithosphere is cracked, somewhat like an eggshell. There are seven or eight

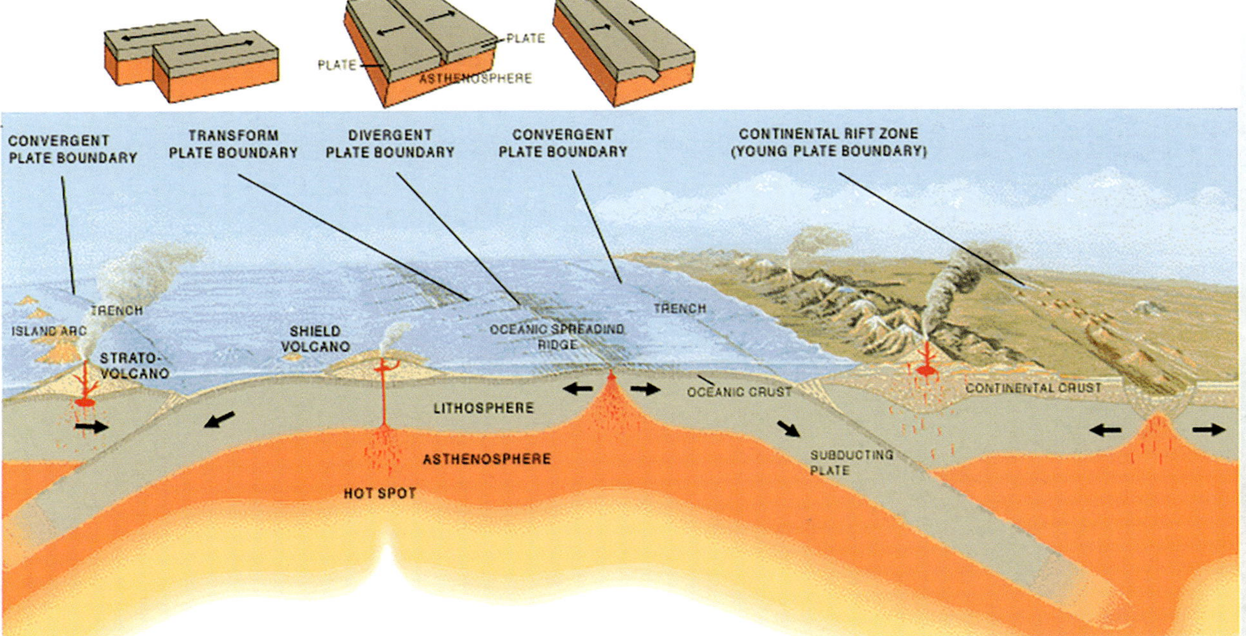

Figure 5.1. Illustration of different tectonic plate boundaries

"Earth's crust and upper mantle together constitute what geophysicists call the *lithosphere*. The lithosphere is cracked, somewhat like an eggshell."

large lithospheric plates and numerous smaller plates. Some of these plates are mostly basaltic oceanic crust. Other plates consist mainly of less dense granitic crust. Today, these plates move very slowly relative to one another, at speeds of about one centimeter per year. Plates move toward each other at *convergent plate boundaries*, while plates move away from one another at *divergent plate boundaries*. Plates slide horizontally past one another at *transform plate boundaries* (Figure 5.1).

Generally, at a convergent plate boundary one of two things will happen. If one plate is oceanic and the other continental, then the more dense oceanic plate will subduct down, sliding under the less dense continental plate. Likewise, an ocean plate can slide under another ocean plate if one of the plates is denser than the other one. If both plates are continental, then they both have roughly the same density, and neither plate subducts below the other. The collision of the plates instead forms a mountain chain.

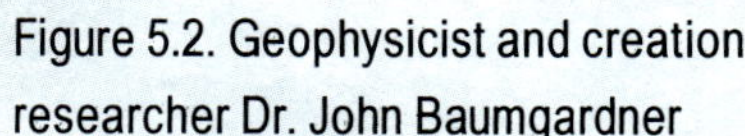

Figure 5.2. Geophysicist and creation researcher Dr. John Baumgardner

As noted earlier, the motions of the plates are very slow. Uniformitarian scientists, consistent with their belief that the present is the key to the past, would argue that the plate motions have always been slow. But is that really the case?

Catastrophic Plate Tectonics (CPT)

One popular news magazine called geophysicist and creation scientist Dr. John Baumgardner (Figure 5.2) "the world's pre-eminent expert" in the use of computers to simulate the movement of material within the earth's mantle.[1] Baumgardner has published much in this area in both secular and creation journals, having studied the field for over 30 years.[2-6] He used sophisticated computer models to simulate the effect of a phenomenon called *runaway subduction* on the motion of the tectonic plates (Figures 5.3 and 5.4). Under the right conditions, plate motions can be thousands of times faster than they are today.

Fluids are characterized by a property called *viscosity*, a measure of internal friction within the fluid. Sticky substance such as maple syrup have high viscosities. The viscosity of silicate rocks in the earth's mantle depends on temperature. Higher temperatures lead to lower viscosities.

Because a colder slab of rock associated with the ocean floor is denser than the surrounding rock, it can sink down into the mantle. When this happens, an "envelope" of material around the subducting slab begins to heat. This heating reduces the viscosity of the surrounding envelope still more, allowing the slab to subduct even faster. This leads to more heating, which leads to an even greater reduction in viscosity, which leads to still more subduction. A runaway effect occurs, which allows the oceanic slab to subduct extremely rapidly down into the lower mantle.

The properties of the mantle allow for this possibility, but this immediately raises a question: Since oceanic plates are still subducting today, then why is catastrophically rapid subduction not still occurring?

One part of the answer is that the density of today's ocean floor prevents it from readily subducting. Also, a competing factor is heat transport within the mantle, which inhibits runaway subduction. Heat generated within the envelope around the subducting slab can be diffused or carried away. Whether or not runaway subduction occurs depends on how quickly this heat

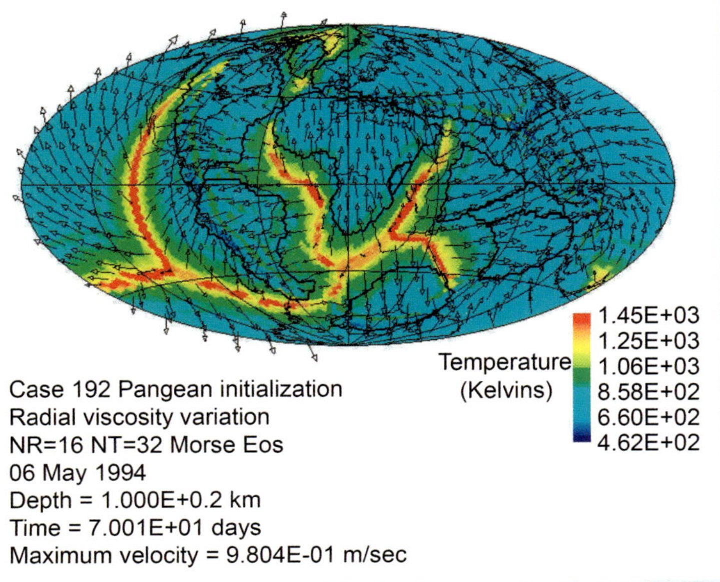

Figure 5.3. Output from one of John Baumgardner's catastrophic plate tectonics (CPT) computer simulations. Figure 4D from reference 6, used by permission of Creation Science Fellowship of Pittsburgh.

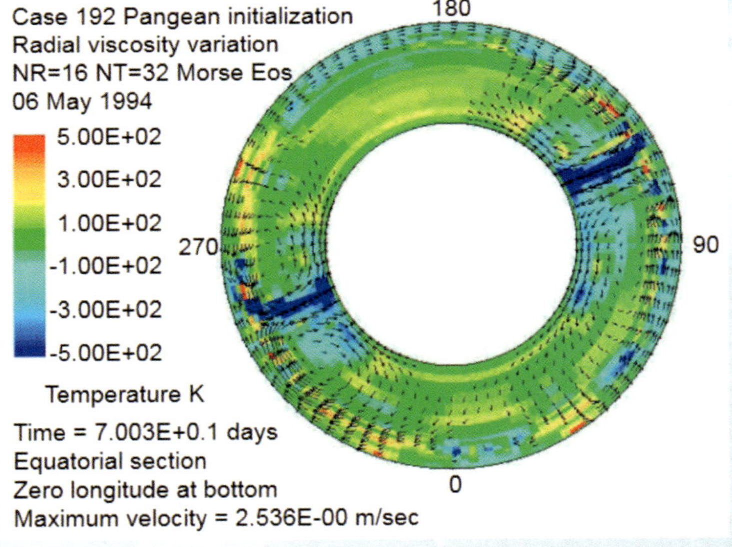

Figure 5.4. Earth's mantle during catastrophically rapid plate subduction, as seen from a slice through the equator. Dark regions at about 120° and 300° are cool, more dense slabs that are subducting down into the warmer, less dense mantle. Figure 5D from reference 6, used by permission of Creation Science Fellowship of Pittsburgh.

(caused by deformation of the mantle material) is carried away from the deforming envelope relative to the speed at which the slab is moving. The runaway effect will only occur if thermal diffusion occurs too slowly to remove this excess heat sufficiently quickly. Conditions within the mantle are close to those needed for such an instability to occur, but fortunately the exact conditions for that do not presently exist.[7] However, a perturbation can nudge the silicate material in the mantle into an unstable state, allowing catastrophically rapid motions of the plates. Baumgardner argues that such a perturbation occurred during the Flood event. He has suggested a number of possible triggers for runaway subduction, including a moderate-size meteorite impact or a single large earthquake.[8]

Suppose that an episode of runaway subduction occurred in the past. What would be the effects? First, subducting ocean slabs would pull the attached continents toward the subduction zones. This would result in rapid, horizontal motions of the plates at speeds of meters per second.[9] Subduction would begin along thousands of miles of continental margins in the pre-Flood world.

“This would result in rapid, horizontal motions of the plates at speeds of meters per second.”

Second, this subduction process would ultimately destroy the pre-Flood ocean floor. And since the earth’s volume remains constant, this destruction of old seafloor means that new seafloor would be created.[10] This seafloor creation occurs at mid-ocean ridges where hot, less dense magma rises. The overlying lithosphere above the ridge stretches and thins, allowing the magma to break through. Today this occurs very slowly, but during the Flood it would have occurred very rapidly along the linear belts stretching thousands of miles along the ocean floor. Baumgardner thinks that as this hot, molten material came into contact with cold seawater, it generated a long, linear geyser of superheated water. Huge amounts of water were ejected into the atmosphere. This water would have then undergone radiative cooling and condensed, causing intense global rain.

Biblical skeptics might scoff at the suggestion that the entire seafloor was destroyed in the Flood, but they should think twice before doing so. Even by secular reckoning, today’s ocean floor is very young. Uniformitarians claim the oldest ocean rock is

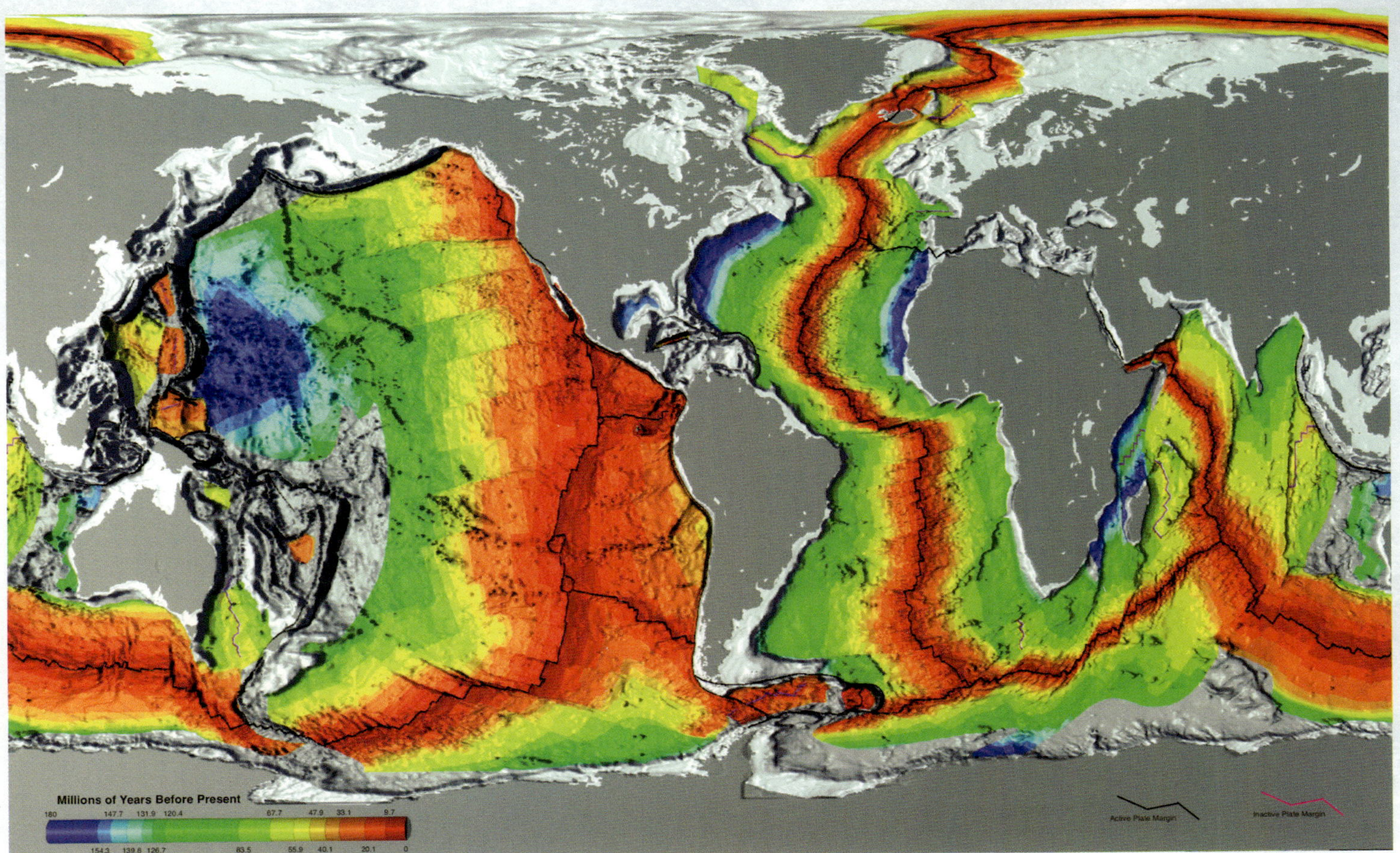

Figure 5.5. Ages (in millions of years) that secular scientists have assigned to different parts of the ocean floor. Both secular and creation scientists agree that the ocean floor is relatively young, although they disagree about the absolute ages.

"only" 200 million years old (Figure 5.5), even though they say the continents and the oceans themselves are billions of years old.[11] So, regardless of whether one believes in a young earth or an old earth, something happened to the previous ocean floor in the relatively recent past. Secular scientists use radioactive dating techniques to calculate these age estimates, but creation scientists think these age estimates are wildly inflated because of many incorrect assumptions regarding radioisotope dating (more on that later).

This upward motion of less dense mantle material at the mid-ocean ridges would have temporarily raised the ocean floor along these linear underwater belts. This would have displaced large amounts of seawater onto the continents, resulting in catastrophic flooding on a global scale. That sounds a lot like the Genesis Flood!

As the newly formed seafloor gradually cooled, it became denser. This would have caused it to sink enough to allow floodwaters to drain off the continents into the ocean basins. The rapidly draining floodwaters would have quickly caused an enormous amount of erosion, forming large, flat planation surfaces on the continents (Figure 5.6). Planation surfaces are not forming today and are very difficult for secular geologists to explain. However, runoff toward the end of the Flood can explain them.[12,13] The fact that seafloor sediment thicknesses are greatest at the continental margins is consistent with enormous erosion late in the Flood.[14,15] In some cases, all the overlying sediments were eroded away, leaving continental bedrock exposed at the surface.

Evidence for CPT

There are many arguments in favor of plate tectonics in general, such as the apparent fit of the different continents, similar fossil distributions on matching parts of the continents, and the increasing thickness of seafloor sediments at greater distances from the mid-ocean ridges.

But there are several reasons to favor the catastrophic version of plate tectonics over the conventional version. If CPT is indeed responsible for much of the geologic upheaval of the Genesis Flood, then these oceanic slabs subducted down into the mantle just 4,500 years ago. Hence, the slabs have had relatively little time to cool. One might expect that these oceanic slabs would be noticeably cooler than the surrounding mantle material.

Seismic tomography is a method in which scientists use waves from earthquakes or explosions to infer variations in composition or temperature of the earth's interior. This method has revealed a ring of dense rock, corresponding approximately to the perimeter of the Pacific Ocean, located at the bottom of the mantle.[16] Inside this ring of material are blobs of less dense rock beneath Africa and the central Pacific. The

Figure 5.6. Cape Breton Highlands planation surface in Nova Scotia, Canada

ring of material lying at the base of the mantle is 3 to 4% denser than these blobs of mantle material. If one assumes this ring has a similar chemical composition to the less dense blob, then this density difference would imply that this ring is 3,000° to 4,000°C cooler than the warmer blob. This is not expected in conventional plate tectonics. The 50 to 100 million years that a slab would take (in uniformitarian thinking) to descend to the base of the mantle should be sufficient time for such temperature differences to even out.[17,18] This strongly suggests that the subduction occurred in the relatively recent past, consistent with an episode of CPT during the Flood. To dodge this conclusion, one would have to attribute these density differences to differences in chemical composition. Some uniformitarian scientists have attempted to make this claim, but Baumgardner considers those arguments contrived.[19,20]

Another Argument for CPT: Rapid Magnetic Reversals

Another argument for CPT comes from the dozens of magnetic reversals recorded in basaltic rocks on either side of the mid-ocean ridges. Molten lava contains iron-laden minerals that tend to align with the direction of the earth's magnetic field at the time the lava cools and solidifies. After it hardens, the direction of the magnetic field is

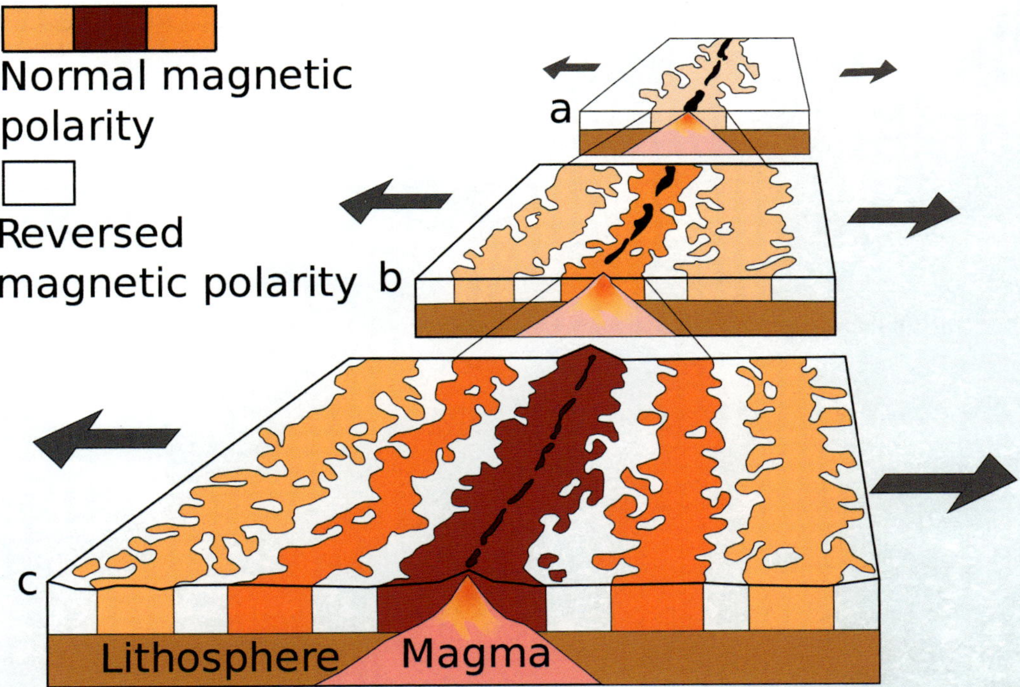

Figure 5.7. Reversals of the direction of the earth's magnetic field are recorded in volcanic rocks on both sides of the mid-ocean ridges

locked into the volcanic rock. The basaltic rocks on either side of the mid-ocean ridges depict a striped pattern like a barcode consisting of alternating bands of magnetization direction that flip back and forth as one moves away from the ridge (Figure 5.7).

If the CPT Flood model is correct, then a new seafloor formed within the year of the Genesis Flood. Since volcanic rocks on this new seafloor show magnetic reversals, that means the magnetic reversals also rapidly occurred during the Flood.

Dr. Russell Humphreys

And we do find evidence that magnetic reversals occurred rapidly. In 1986, creation scientist and physicist Russell Humphreys proposed a way to test the hypothesis of fast magnetic reversals. Thin lava flows only need a few weeks to cool and solidify. If a thin lava flow is found that records a full reversal of the earth's magnetic field, then it would be strong evidence that the magnetic reversal could have taken no longer than a couple of weeks—the time it took for the lava to cool and crystallize.[21]

One year before Humphreys made his prediction, four uniformitarian scientists reported evidence of extremely rapid changes in Earth's magnetic field.[22-24] Most uniformitarian scientists had great difficulty believing the results. However, scientists presented additional evidence for rapid magnetic reversals over the next 10 years.[25,26] In fact, scientists estimated that the direction of the earth's magnetic field was changing as much as 6° per day for 13 consecutive days—twice the originally reported rate.[27,28] One of the original scientists later repudiated the results, but other scientists have since also claimed evidence for rapid changes in the earth's magnetic field.[29-32]

So, at least some reversals of the earth's geomagnetic field occurred rapidly. But if some reversals occurred rapidly, then they probably all occurred rapidly. The reason is

"In fact, uniformitarian scientists do not even have a working theory that explains how the earth can have a magnetic field that has persisted for billions of years."

that explaining a magnetic reversal is quite difficult. In fact, uniformitarian scientists do not even have a working theory that explains how the earth can have a magnetic field that has persisted for billions of years.[33,34] Hence, they have trouble explaining the very existence of the earth's magnetic field, let alone the reversals that have occurred in the past.

Given the theoretical difficulties in explaining a magnetic reversal, a single mechanism for all the reversals is more likely than multiple mechanisms. And if that single mechanism caused some of the reversals to occur quickly, then it probably caused all of them to occur quickly. And since these rapid magnetic reversals are recorded in volcanic rocks on the ocean floor, this implies that the creation of new seafloor was also rapid. But, as noted earlier, creation of a new seafloor implies destruction of the old seafloor. So, the destruction of the old seafloor must also have been rapid. But rapid destruction of old seafloor implies rapid subduction, as expected in CPT and possibly confirmed by seismic tomography (Figure 5.8).

Humphreys has also developed a theory that explains, at least conceptually, how such rapid magnetic reversals could occur.[35] His mechanism requires strong up-and-

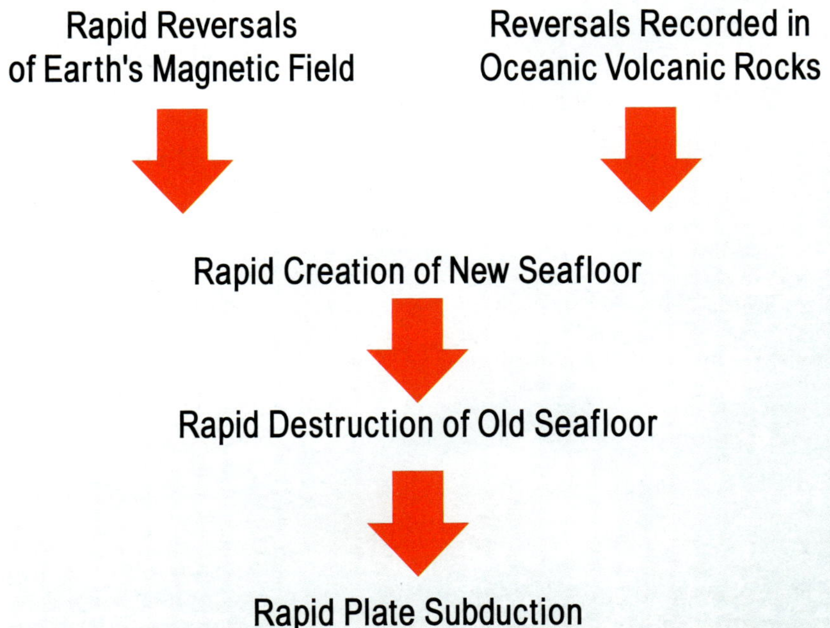

Figure 5.8. Even some secular scientists now acknowledge that at least some past reversals of the earth's magnetic field have been extraordinarily fast. If most or all of the reversals occurred quickly, then the fact that these rapid reversals are recorded in oceanic volcanic rocks implies the rapid creation of a new seafloor and an accompanying rapid destruction of the old seafloor via rapid plate subduction, as predicted by the theory of catastrophic plate tectonics.

down motion of fluids within the earth's outer liquid core due to convection. This convection could start if a cold subducting plate were to come into contact with the hot outer liquid core at the core-mantle boundary.[36] Of course, this is exactly what is expected in CPT.

To summarize, if a relatively cold, dense oceanic plate really did rapidly subduct down into the mantle, as expected in CPT, then one would expect rapid convection in the outer core. This rapid convective motion would cause rapid reversals of the earth's magnetic field, as indicated by the field evidence. So, contrary to popular misperception, plate tectonics and reversals of the earth's magnetic field are not problems for the creation/Flood model. Rather, a CPT Flood model explains them handily.

Too Much Heat?

Why spend so much time on catastrophic plate tectonics in a book about climate change? There is a good reason. When hot magma rapidly emerged from the mid-ocean ridges during the Genesis Flood, this hot magma rapidly heated the surrounding oceanwater. Rapid subduction would also cause an enormous number of explosive volcanic eruptions. This is because earthquakes and explosive volcanic eruptions are associated with subducting plates. As it turns out, both the heating of the world's oceans and intense volcanism are critically important in explaining the Ice Age.

However, although this scenario makes a great deal of sense, there is an apparent problem with it. The amount of heat generated during the Flood would have been enormous, sufficient to melt the crust several times over and wipe out all life on Earth. How could this heat be safely dissipated? Some of this heat is critical to explain an Ice Age, but too much heat is a very real problem. The correct solution to this apparent problem will likely require some kind of divine intervention, or tweaking, of the laws of physics; i.e., a miracle.[37] In fact, Humphreys has already proposed some possible economic miracles to dissipate the excess heat, the most recent of which is based on multiple passages of Scripture discussing the "opening" of the heavens.[38,39]

Of course, the skeptic will likely dismiss the possibility that God may have altered the laws of physics during the Flood. However, this response is somewhat inconsistent since secular scientists have also suggested, for various reasons, past changes in the laws of physics.[40-43] Creation scientists readily acknowledge both God's right and ability to intervene in history. We also acknowledge the historical evidence that God has supernaturally intervened in history. Of course, these interventions in history are relatively rare, as suggested even by Scripture (1 Samuel 3:1).[44] God is not the author of confusion (1 Corinthians 14:33), and He does not overrule the laws that He himself established without good reason. Hence, creation scientists still expect an underlying orderliness in nature, an orderliness that makes science possible.

Creation scientists would offer the following arguments that God supernaturally intervened in order to safely dissipate excess heat during the Flood. First, there is overwhelming physical and historical evidence that the Flood did occur. Since this is the case, then God somehow must have safely dissipated this excess heat. Second, warm post-Flood oceans caused by this heating enable creation scientists to solve multiple mysteries of Earth history, histories that still perplex secular scientists. We will discuss a number of these mysteries later. Third, there is already strong evidence that God has altered the laws of physics. Creation scientists have found multiple lines of evidence that nuclear decay rates have varied in the past.[45] Such an acceleration demands a tweaking of the laws of physics.

Accelerated Nuclear Decay

But accelerated nuclear decay raises other questions. When did this accelerated decay occur, and why would God do it? The fact that radiation is harmful to living things puts restrictions on when accelerated decay could occur and how much could occur at a given time. Accelerated nuclear decay could plausibly occur at several times.[46]

1. It could have occurred before the end of Day 3 of the creation week, before God created living things that could be harmed by radioactivity.
2. During the Flood, tectonic and geologic upheaval allowed radioactive elements in the crust to mix with the floodwaters. These radioactive elements would have been potentially harmful to Noah and the other inhabitants on the Ark, but the very deep water (the average water depth could easily have been a mile or so) and thick wooden walls of the Ark would have offered them some protection from these radioactive elements.
3. Such decay could have happened at the time of the Fall of humanity if the radioactive elements remained deep within the earth's crust where they could not harm living things.

"Then, at the time of the Flood God accelerated this nuclear decay, and the extra heat from this accelerated decay may have been the 'trigger' that started the geologic upheaval of the Flood."

But why would God put radioactivity, which can harm living things, in a "very good" creation (Genesis 1:31)? And why would He speed up this radioactivity? First, radioactivity may have been completely absent in the completed "very good" pre-Fall world. It would certainly have been absent from the biosphere. Some have suggested that God may have caused radioactivity as part of His cursing of the ground in Genesis 3:17 (see also Deuteronomy 32:22).[47,48] In that case, the rates of radioactive decay were steady for the 1,656 years between the Fall and the Flood. Then, at the time of the Flood God accelerated this nuclear decay, and the extra heat from this accelerated decay may have been the "trigger" that started the geologic upheaval of the Flood. In both cases, God would have been using radioactivity for a purpose. His tweaking of the laws of physics was not arbitrary.

Of course, changing nuclear decay constants would completely invalidate the age estimates of millions of years obtained from radioisotope dating methods.[49] In addition, proposed mechanisms for accelerated decay would not necessarily affect the chemistry upon which organic life depends since this chemistry depends on the arrangements of electrons within atoms, not the properties of the nucleus per se.[50]

The heat generated during the Flood plays a critical role in explaining the Ice Age, although the amount of heat needs to be dialed back somewhat. As we shall see, uniformitarian Ice Age theories have the opposite problem—their proposed mechanisms are much too weak to cause an Ice Age.

References

1. Chandler, B. 1997. The Geophysics of God. *U.S. News & World Report.* (June 16): 55-58.
2. Baumgardner, J. R. 1985. Three-dimensional treatment of convective flow in the earth's mantle. *Journal of Statistical Physics.* 39 (5): 501-511.
3. Baumgardner, J. R. 1993. 3-D numerical investigation of the mantle dynamics associated with the breakup of Pangea. In *Flow and Creep in the Solar System: Observations, Modeling, and Theory.* D. B. Stone and S. K. Runcorn, eds. NATO ASI Series. 391: 207-224.
4. Baumgardner, J. R. 1986. Application of supercomputers to 3-D mantle convection. In *The Physics of the Planets.* S. K. Runcorn, ed. New York: John Wiley & Sons, 199-231.
5. Baumgardner, J. R. 1994. Runaway Subduction as the Driving Mechanism for the Genesis Flood. In *Proceedings of the Third International Conference on Creationism.* R. E. Walsh, ed. Pittsburgh, PA: Creation Science Fellowship, 63-75.
6. Baumgardner, J. R., Computer Modeling of the Large-Scale Tectonics Associated with the Genesis Flood, ibid, 49-62.
7. Baumgardner, J. R. 1990. 3-D Finite Element Simulation of the Global Tectonic Changes Accompanying Noah's Flood. In *Proceedings of the Second International Conference on Creationism.* R. E. Walsh and C. L. Brooks, eds. Pittsburgh, PA: Creation Science Fellowship, 35-45.
8. Baumgardner, Runaway Subduction as a Driving Mechanism for the Genesis Flood, 74.
9. Austin, S. A. et al. 1994. Catastrophic Plate Tectonics: A Global Flood Model of Earth History. In *Proceedings of the Third International Conference on Creationism.* R. E. Walsh, ed. Pittsburgh, PA: Creation Science Fellowship, 609-621.
10. Van Andel, T. H. and J. B. Murphy. 2015. Plate Tectonics. *Encyclopaedia Britannica Online.* Posted on britannica.com, accessed December 5, 2016.
11. Luyendyk, B. P. Oceanic crust. *Encyclopaedia Britannica Online.* Posted on britannica.com, accessed June 3, 2014.
12. Twidale, C. R. 1982. Planation surfaces—Theory and reality. In *Perspectives in Geomorphology*, vol. 1. H. S. Sharma, ed. New Delhi, India: Naurang Rai, Concept Publishing Co, 63-82.
13. Oard, M. J. 2011. The remarkable African Planation Surface. *Journal of Creation.* 25 (1): 111-122.
14. Deep-sea Sediments. *Encyclopaedia Britannica online.* Posted on britannica.com, accessed February 21, 2019.
15. Clarey, T. 2015. The Whopper Sand. *Acts & Facts.* 44 (3): 14.
16. Su, W.-J., R. L. Woodard, and A. M. Dziewonski. 1994. Degree-12 model of shear velocity heterogeneity in the mantle. *Journal of Geophysical Research.* 99: 6945-6980.
17. Baumgardner, J. R. Noah's Flood: The Key to the Correct Interpretation of Earth History. Southern California Seminary. Posted on socalsem.edu September 18, 2013 accessed June 5, 2017.
18. Baumgardner, J. R. 2003. Catastrophic Plate Tectonics: The Physics Behind the Genesis Flood. In *Proceedings of the Fifth International Conference on Creationism.* R. L. Ivey, Jr., ed. Pittsburgh, PA: Creation Science Fellowship, 113-126.
19. Kellogg, L. H., B. H. Hager, and R. D. van der Hilst. 1999. Compositional Stratification in the Deep Mantle. *Science.* 283: 1881-1884.
20. Baumgardner, Noah's Flood: The Key to the Correct Interpretation of Earth History.
21. Humphreys, D. R. 1986. Reversals of the Earth's Magnetic Field During the Genesis Flood. In *Proceedings of the First International Conference on Creationism.* R. L. Ivey, Jr., ed. Pittsburgh, PA: Creation Science Fellowship, 113-126.
22. Prévot, M. et al. 1985. How the geomagnetic field vector reverses polarity. *Nature.* 316: 230-234.
23. Mankinen, E. A. et al. 1985. The Steens Mountain (Oregon) geomagnetic polarity transition, 1. Direction variation, duration of episodes, and rock magnetism. *Journal of Geophysical Research.* 90: 10, 393-10, 416.
24. Prévot, M. et al. 1985. The Steens Mountain (Oregon) geomagnetic polarity transition, 2. Field intensity variations and discussion of reversal models. *Journal of Geophysical Research.* 90: 10, 417-10, 448.
25. Coe, R. S. and M. Prévot. 1989. Evidence suggesting extremely rapid field variation during a geomagnetic reversal. *Earth and Planetary Science Letters.* 92: 292-298.
26. Coe, R. S., M. Prévot, and P. Camps. 1995. New evidence for extraordinarily rapid change of the geomagnetic field during a reversal. *Nature.* 374: 687-692.
27. Merrill, R. T. 1995. Principle of least astonishment. *Nature.* 374: 674-675.
28. Bergeron, L. When north flies south. *New Scientist.* Posted on newscientist.com March 30, 1996, accessed December 7, 2016.
29. Coe, R. S. et al. 2014. Demise of the rapid-field-change hypothesis at Steens Mountain: The Crucial Role of Continuous Thermal Demagnetization. *Earth and Planetary Science Letters.* 400: 302-312.
30. One cannot help but wonder if political pressure may have played a role in this repudiation since creation scientists have made much of the evidence for rapid geomagnetic reversals over the years.

31. Bogue, S. W. and J. M. G. Glen. 2010. Very rapid geomagnetic field change recorded by the partial remagnetization of a lava flow. *Geophysical Research Letters.* 37 (21): L21308.
32. Sagnotti, L. et al. 2014. Extremely rapid directional change during Matuyama-Brunhes geomagnetic polarity reversal. *Geophysics Journal International.* 199 (2): 1110-1124.
33. Humphreys, D. R. 2013. Planetary Magnetic Dynamo Theories: A Century of Failure. In *Proceedings of the Seventh International Conference on Creationism*. M. Horstemeyer, ed. Pittsburgh, PA: Creation Science Fellowship.
34. Folger, T. Journeys to the Center of the Earth: Our planet's core powers a magnetic field that shields us from a hostile cosmos. But how does it really work? *Discover*. Posted on discovermagazine.com July 13, 2014, accessed January 18, 2017.
35. Humphreys, D. R. 1990. Physical Mechanism for Reversals of the Earth's Geomagnetic Field During the Flood. In *Proceedings of the Second International Conference on Creationism*. R. E. Walsh and C. L. Brooks, eds. Pittsburgh, PA: Creation Science Fellowship, 129-142.
36. Austin et al., Catastrophic Plate Tectonics, 612.
37. Baumgardner, J. R. 1986. Numerical Simulation of the Large-Scale Tectonic Changes Accompanying the Flood. In *Proceedings of the First International Conference on Creationism*. R. E. Walsh, C. L. Brooks, and R. S. Crowell, eds. Pittsburgh, PA: Creation Science Fellowship, 17-28.
38. Humphreys, D. R. 2000. Accelerated Nuclear Decay: A Viable Hypothesis? In *Radioisotopes and the Age of the Earth*, vol. 1. L. Vardiman, A. A. Snelling, and E. F. Chaffin, eds. El Cajon, CA and St. Joseph, MO: Institute for Creation Research and the Creation Research Society, 333-379.
39. Humphreys, D. R. 2018. New Mechanism for Accelerated Removal of Excess Radiogenic Heat. In *Proceedings of the Eighth International Conference on Creationism*. J. H. Whitmore, ed. Pittsburgh, PA: Creation Science Fellowship, 731-739.
40. Becker, K. Are the Laws of Physics Really Universal? PBS. Posted on pbs.org October 21, 2015, accessed January 19, 2017.
41. Collins, D. Has the Speed of Light Slowed Down? *CBS News*. Posted on cbsnews.com August 7, 2002, accessed January 19, 2017.
42. Ball, P. Can the laws of physics change? BBC. Posted on bbc.com March 29, 2012, accessed January 19, 2017.
43. Davies, P. 2007. Laying Down the Laws. *New Scientist.* 194 (2610): 30-34.
44. Given God's great interest and love for us, it would not be at all surprising if genuine miracles are fairly common. Indeed, Christians can testify to many instances in which God has obviously performed supernatural deeds (supernatural healing, miraculous provisions, etc.). But because these miracles are distributed among billions of people, they seem quite rare from the perspective of a single person.
45. Vardiman, L. et al. 2003. Radioisotopes and the Age of the Earth. In *Proceedings of the Fifth International Conference on Creationism*. R. L. Ivey Jr., ed. Pittsburgh, PA: Creation Science Fellowship, 337-348.
46. Humphreys, Accelerated Nuclear Decay, 340-341.
47. Johnson, J. J. S., personal communication, fall 2018.
48. Humphreys, Accelerated Nuclear Decay, 356-357.
49. Vardiman et al., *Radioisotopes and the Age of the Earth*.
50. Humphreys, Accelerated Nuclear Decay, 365.

6 The World That Then Was

Summary: God's original creation was "very good," without any death or suffering either of people or animals. Earth's original climate and biosphere were also very good and likely remained that way even after the Fall, generally temperate with abundant vegetation. A temperate pre-Flood climate with abundant vegetation is consistent with the fossil evidence, although uniformitarian scientists have mistakenly assigned ages of millions of years to sedimentary rock units that were actually formed during the Flood. An interesting point directly relevant to the current climate change discussion is that today's concentration of atmospheric carbon dioxide is about 410 ppm (parts per million). In other words, 410 of every 1,000,000 gas molecules in the atmosphere are carbon dioxide molecules. It's worth noting that secular scientists think that 500 million years ago the amount of atmospheric CO_2 was 10 to 20 times higher than even this supposedly very high value, and they think past climates were warmer as well. So, even by secular reckoning today's CO_2 values and temperatures are not unprecedented. The next chapter will discuss errors in secular reasoning regarding climate change.

In a discussion of Genesis and climate, it is certainly appropriate to consider the climate of both the pre-Fall and pre-Flood worlds. Unless one has specific reasons for assuming that a climatic change occurred at the time of Adam's sin, it seems likely to me that the pre-Flood climate was basically the same as the pre-Fall climate. Clearly, God allowed bad things such as thorns and thistles, as well as death and suffering, into the world at that time. However, there doesn't seem to be an obvious reason to assume that the climate also changed. Because "the world that then existed perished, being flooded with water" (2 Peter 3:6), we no longer have direct access to that world.[1]

However, there are clues in the first chapters of Genesis that are relevant to a biblical consideration of climate.

A "Very Good" World

First, God's original creation was "very good" (Genesis 1:31), without death (Figure 6.1) or suffering (Romans 5:12; 8:20-22). Christians who accept the doctrine of an old earth almost invariably attempt to argue that animal death was in the world prior to Adam's sin, even though some correctly claim that human death was absent. They do so because they accept the claim that fossils are the remains of animals that lived and died millions of years before the appearance of humans.

However, the whole tenor of Scripture argues against this. First, there is no hint of millions of years anywhere in the Bible. It's just not there! Second, Scripture teaches us that "the LORD is good to all: and His tender mercies are over all His works" (Psalm 145:9). Since God cares for His creatures, why would He allow millions of years of animal death and suffering? Even we humans, evil though we are (Matthew 7:11), can be moved with pity at the sight of a dead dog on the side of the road. If God indifferently looked on millions of years of animal death and suffering, what would this say about God's character? What would it say about His character if He looked on with apparent

Figure 6.1. Both animal and human death were absent in God's originally "very good" creation (Genesis 1:31)

> "It is precisely because death and suffering were absent in the pre-Fall world that this shedding of blood is so significant."

indifference for hundreds of millions of years while animals experienced bloodshed, disease, and death—including violent death?

No, God's creation was very good. Both people and animals were initially vegetarians (Genesis 1:29-30). This means there was no bloodshed in the pre-Fall world. Christians who advocate an old earth are forced to conclude that animal bloodshed existed before the Fall. In doing so, they unintentionally minimize the significance of the animal sacrifices God instituted at the Fall (Genesis 3:21; 4:4-5). It is precisely because death and suffering were absent in the pre-Fall world that this shedding of blood is so significant. Animal slaughter is horrific, and people often become faint or queasy when first exposed to the sight of it—say, on a farm. But its horrific nature *is precisely the point*. Rebellion against God is horrific and demands a horrific penalty, the shedding of blood (Hebrews 9:22). However, the presence of bloodshed in the world prior to Adam's sin minimizes both the horror and significance of animal sacrifice, which foreshadowed Christ's shedding of His own blood on the cross. If there was no shedding of blood, then clearly Adam and Eve did not eat meat. Nor did the animals before the Fall. Of course, that makes perfect sense in light of the fact that both people and animals were originally vegetarians.

Some environmentalists and animal rights activists strongly oppose the eating of meat even though God has, since the time of Noah, allowed us to eat it (Genesis 9:2-4; 1 Timothy 4:1-5). Although they err in condemning the eating of animal meat, this error is rooted in the correct realization that animal death is bad. Yes, God has given us permission to eat animal meat (possibly to supplement our diets with nutrients that became harder to obtain after the Flood), but it was not always so. And animals will no longer eat one another when Christ returns to set up His kingdom (Isaiah 11:6; 65:25). It may be that humans will no longer eat meat at that time as well.

However, opposition to the eating of meat, though misguided, makes sense in a biblical worldview. This is because Christianity sees death as unnatural and an "enemy" (1 Corinthians 15:26). In a Darwinian worldview, animals have been slaughtering and eating each other for millions of years. So, why is it wrong for humans to kill and eat less-evolved creatures? Furthermore, from an evolutionary perspective, how can one say that any action is right or wrong?

A "Very Good" Climate

God described His finished creation as "very good" (Genesis 1:31). Not surprising-

Figure 6.2. The pre-Flood world had lush, abundant vegetation

ly, the fossils show that there was abundant vegetation in the "very good" pre-Flood world (Figure 6.2). I suspect that there were no deserts or ice sheets, but I don't want to be dogmatic about that. By the way, we can now shed some light on Dr. Eric Pianka's counterintuitive belief that "greening" of the deserts is a bad thing (see chapter 2). In the biblical view, God created an optimal "very good" world with abundant vegetation. In the evolutionary view, however, there was no primeval paradise. Harsh environments in which animals struggle to eke out an existence have always existed. In this scenario, deserts and polar ice sheets are the natural state of things. Since organisms supposedly evolved to fill those particular environmental niches, Pianka assumes these creatures will necessarily have difficulty surviving in new environments. This is true even if the new environments are apparently "better" (with more vegetation) than the old ones! Pianka apparently doesn't think that these creatures can live in environments with abundant vegetation because it was these very desert environments, through an evolutionary process, that supposedly *created* these animals.

But God, not the environment, created living things. So, we shouldn't be surprised that creation scientists are finding abundant evidence that creatures have a God-given ability to rapidly adjust to new environmental conditions, including those such as deserts that might be less than ideal.[2-4] Deserts and ice sheets might not have existed at all in the pre-Flood world, but if they did exist they were probably much smaller than they are now. So, it is not surprising that Christ will cause the deserts to bloom when He returns.

> Because the palaces will be forsaken, the bustling city will be deserted. The forts and towers will become lairs forever, a joy of wild donkeys, a pasture of flocks—until the Spirit is poured upon us from on high, and the wilderness becomes a fruitful field, and the fruitful field is counted as a forest. (Isaiah 32:14-15)

> The wilderness and the wasteland shall be glad for them, and the desert shall rejoice and blossom as the rose; It shall blossom abundantly and rejoice, Even with joy and singing. The glory of Lebanon shall be given to it, the excellence of Carmel and Sharon. They shall see the glory of the LORD, the excellency of our God....Then the eyes of the blind shall be opened, and the ears of the deaf shall be unstopped. Then the lame shall leap like a deer, and the tongue of the dumb sing. For waters shall burst forth in the wilderness, and streams in the desert. The parched ground shall become a pool, and the thirsty land springs of water; in the habitation of jackals, where each lay, there shall be grass with reeds and rushes. (Isaiah 35:1-2, 5-7)

But this raises a question. If the world was originally a primeval paradise with abundant vegetation, then how and when did harsher conditions (including large deserts and large ice sheets) come about? We will answer that question later.

For waters shall burst forth in the wilderness, and streams in the desert. (Isaiah 35:6)

The Fossil Evidence for a Global Flood

A temperate global pre-Flood world with much more abundant vegetation is consistent with the fossil evidence. As noted earlier, secular scientists claim that the oldest seafloor on the planet is 200 million years old. This corresponds, by their reckoning, with the beginning of the Jurassic period.[5] In the creation model, this would be the beginning of the most catastrophic part of the Flood, with its accompanying rapid seafloor spreading and catastrophic plate motions.

In this light, it's worth noting that secular scientists believe that the Jurassic and Cretaceous periods saw "extreme climatic warmth, massive organic-carbon burial and carbon-cycle disruption, near-global anoxia in the world's oceans, and major biotic crises."[6] Of course, if one strips away the fictitious old-earth doctrine, this sounds a lot like the Flood! Clearly, massive organic-carbon burial did occur during the Flood. Creation scientists think the pre-Flood world contained much more carbon than is present in today's world, much of which was sequestered during the Flood event.[7]

Even in the polar regions, scientists have found fossils of plants and animals that preferred a warmer climate.[8,9] Either these creatures lived at high latitudes before the

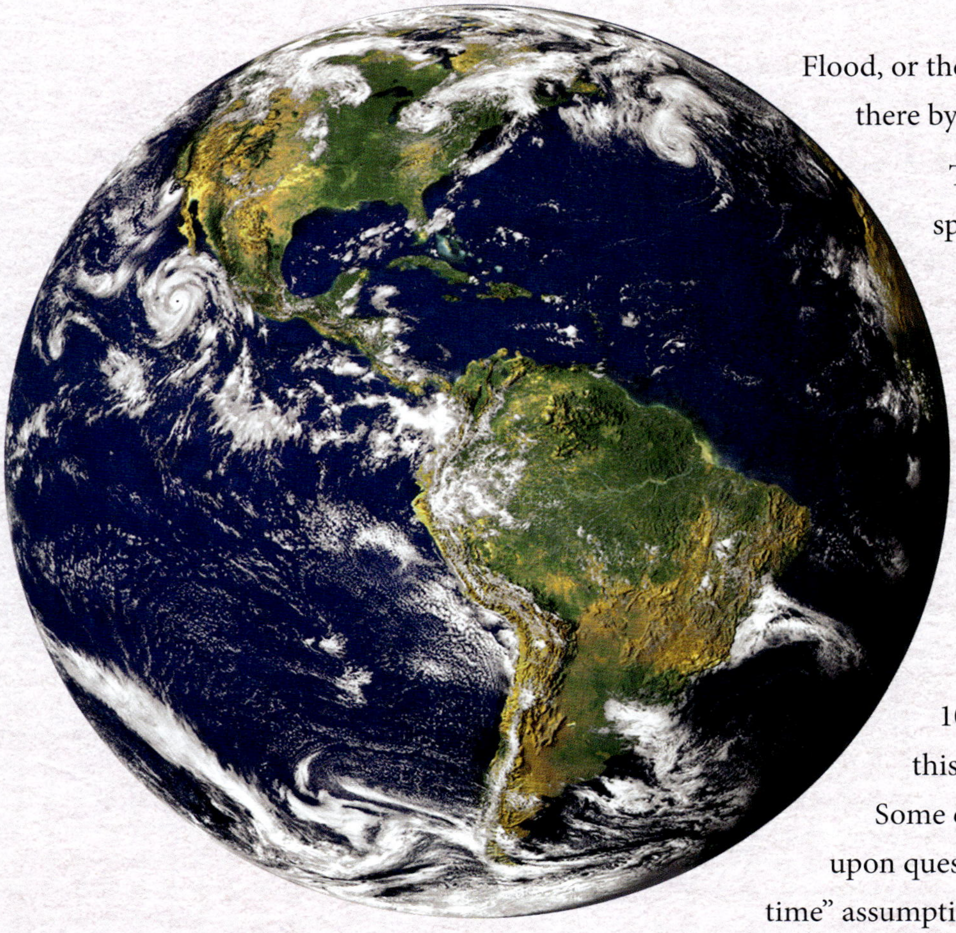

Flood, or their remains were transported there by Flood currents.

Today's concentration of atmospheric carbon dioxide is about 410 ppm (parts per million). This means that in a sample of dry air (having no water vapor) 410 of every 1,000,000 gas molecules are carbon dioxide molecules. It's worth noting that secular scientists think that 500 million years ago the amount of atmospheric CO_2 was 10 to 20 times higher than even this supposedly very high value.[10,11] Some of these estimates might be based upon questionable evolutionary "deep time" assumptions, but carbon dioxide levels probably were higher in the past.[12,13] In fact, since CO_2 is a greenhouse gas, it may have been a contributing factor to the temperate climate existing in the pre-Flood world.[14] This helps provide some perspective on rising CO_2 levels.

But if secular scientists believe that atmospheric CO_2 levels in the past were so much higher, then why are they so worried about today's atmospheric CO_2, which is much lower by comparison? Remember Dr. Katharine Hayhoe's concern about past CO_2 levels:

> We are so far outside the range of natural variability. We have not seen carbon dioxide levels like this in the history of human civilization on this planet.[15]

Geologist Howard Lee sheds some light on why secular scientists are concerned:

> Life flourished in the Eocene, the Cretaceous and other [prehistoric] times of high CO_2 in the atmosphere because the greenhouse gasses

> were *in balance* with the carbon in the oceans and the weathering of rocks. Life, ocean chemistry, and atmospheric gasses *had millions of years to adjust* to those levels.[16]

So, apparently it isn't the change in CO_2 or the amount of warming that has uniformitarian scientists so concerned. Rather, it's the supposed speed of the change. Lee says there have been abrupt warming events in the prehistoric past but that these events were harmful to living things:

> Those abrupt global warming events were almost always highly destructive for life, causing mass extinctions such as at the end of the Permian, Triassic, or even mid-Cambrian periods....In all cases [of past climate change] we see the same association between CO_2 levels and global temperatures. And past examples of *rapid* carbon emissions (just like today) were generally highly destructive to life on Earth.[17]

These conclusions are based on an evolutionary, old-earth interpretation of the rocks and fossils. But these rocks were really deposited during the Genesis Flood. Was the Flood highly destructive to life on Earth? Of course it was! Did carbon dioxide levels change during the Flood? Certainly—probably dramatically. In fact, because the Flood lasted only a year, the changes were much faster than even secular scientists claim. But it was the Flood itself, not the accompanying changes in carbon dioxide, that caused this destruction. Secular scientists are attributing the devastation that occurred during the Flood to higher CO_2 levels when the higher CO_2 levels were a *consequence* of the devastation, not the cause.

"But it was the Flood itself, not the accompanying changes in carbon dioxide, that caused this destruction."

Yet, life survived and quickly adjusted to the new post-Flood CO_2 levels. Why? Because God, in His wisdom and foreknowledge, equipped these creatures with the genetic and epigenetic tools to quickly respond to a changing environment (more on this later). And don't forget that CO_2 levels were probably higher in the pre-Flood world. Remember also that this higher amount of CO_2 corresponded with a temperate, optimal climate. After all, a thriving, healthy (albeit very sinful) human population was present at this time—people were still living to be 900 years old before the Flood.

Creationists don't think the extreme longevity of pre-Flood humans was solely the result of higher CO_2 levels. In fact, higher CO_2 levels probably had little, if anything, to do with it. Most creationists think a much more pristine genome, with a small number of harmful mutations, was the main cause of these long life spans, although a much more ideal environment may also have contributed.[18,19]

However, this does give us some perspective on the global warming issue. If God's original "very good" world had an atmosphere with much higher CO_2 levels, then is an increase in atmospheric CO_2 something over which we should be panicking? The current amount of atmospheric carbon dioxide is probably not the highest it has ever been. Even the rate of change is likely not the highest it has ever been. This illustrates the point I made earlier—what you believe about Genesis will profoundly affect the way you view both past and future climate change.

An Aside: Living to Be More Than 900?

It is worthwhile to briefly consider the extremely long life spans recorded in the earliest chapters of Genesis. These life spans have long been the subject of ridicule by the skeptics. As noted earlier, most creation scientists think that accumulated genetic defects are the primary cause of our much shorter life spans. It's an empirical fact that our genomes are rapidly deteriorating, with hundreds of new mistakes occurring in each generation. If these mistakes were not present, would our life spans be much greater than they are now? If you think about it, we don't really know that a 70-year life span is normal. We just think it's normal because we have gotten used to it over the centuries.

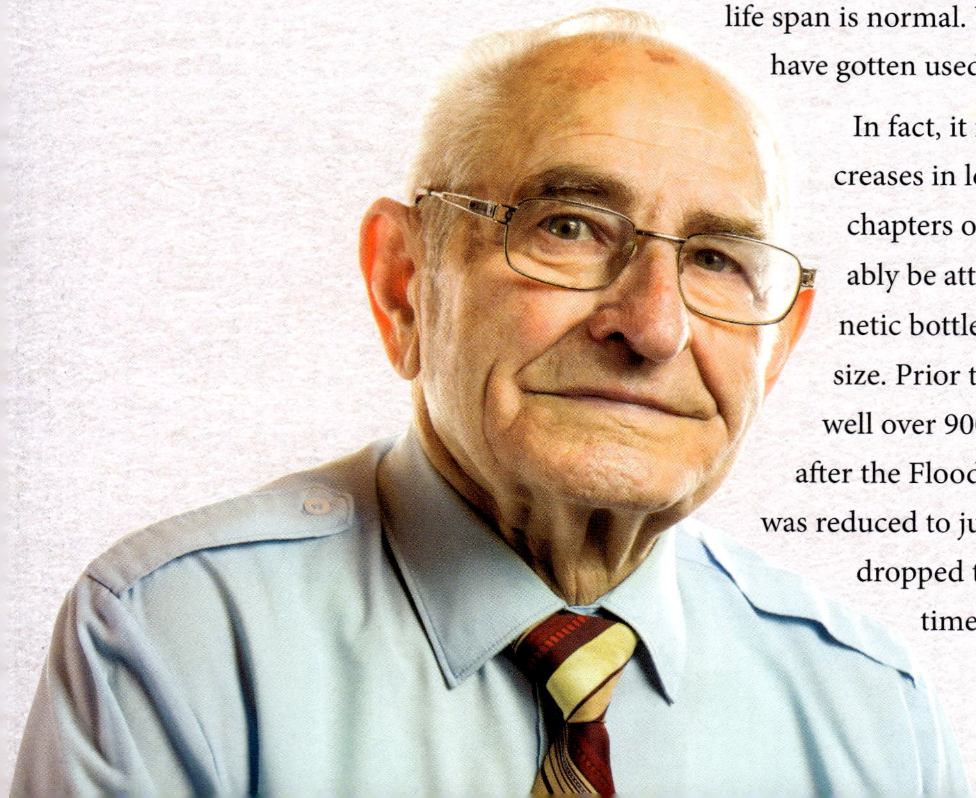

In fact, it is striking that two dramatic decreases in longevity are recorded in the earliest chapters of Genesis, decreases that can reasonably be attributed to the harmful effects of "genetic bottlenecks," or large drops in population size. Prior to the Flood, most people lived to be well over 900 years old (Genesis 5). However, after the Flood, and after the human population was reduced to just eight people, the typical life span dropped to about 400 to 500 years until the time of Peleg (Genesis 11:10-17), where it dropped again. Many creationists think Peleg lived around or shortly after the Tower of Babel.[20] Since the human population was dispersed from the Tower of Babel, then more genetic bottlenecks would have occurred at this time as the families scattered. Is this dramatic drop in longevity at the time of Peleg a coincidence?

> "It's an empirical fact that our genomes are rapidly deteriorating, with hundreds of new mistakes occurring in each generation."

No Rain Before the Flood?

There are hints in Genesis that a different hydrological cycle operated in the pre-Flood world. Based on Genesis 2:4-6, many creation scientists think rain was absent.

> This is the history of the heavens and the earth when they were created, in the day that the LORD God made the earth and the heavens, before any plant of the field was in the earth and before any herb of the field had grown. For the LORD God had not caused it to rain on the earth, and there was no man to till the ground; but a mist went up from the earth and watered the whole face of the ground. (Genesis 2:4-6)

Obviously, an absence of rain in the pre-Flood world requires explanation. However, other creationists argue, with some plausibility, that the absence of rain was confined just to the creation week itself prior to humanity's creation (Genesis 2:5). In that case, no explanation would be necessary.

However, there are several good reasons to conclude that rain really was absent in the pre-Flood world, or at least absent from Earth's surface. First, if the absence of rain was confined to just the first five days of the creation week, then why mention it at all? It is perfectly common in today's world, and hardly noteworthy, for rain to be absent in a given area for five days or more. The fact that this absence of rain is mentioned seems to suggest that the Holy Spirit (2 Timothy 3:16) considered it noteworthy and worth mentioning.

Likewise, the passage contrasts rain with an alternate watering mechanism: "a mist went up from the earth and watered the whole face of the ground" (Genesis 2:6). Is

Clouds seem to have been present in the pre-Flood world even if rain was not

there any reason to assume that this mist ceased once God created Adam? This seems unlikely, especially since the mist was serving the useful and needed purpose of watering the face of the ground.

Also, if one assumes there was rain in the pre-Flood world, then this raises the possibility of torrential downpours. The Bible makes no mention of a house for Adam and Eve in the Garden of Eden. The temperate climate presumably allowed them to sleep under the stars. But without a house, Adam and Eve would have been exposed to rainfall, including possibly heavy rainfall. I have a hard time imagining Adam and Eve enjoying an hours-long downpour, even if it posed no danger to them.

For these reasons, I tend to think the traditional creationist understanding of this passage is correct. Rain was indeed absent from the pre-Flood world, or at least from the land's surface. Note that the text says that the rain did not fall on the "earth" or the "whole face of the ground," which leaves open the possibility that rain may still have fallen offshore.However, clouds apparently *were* present. In Proverbs 8, Solomon personifies wisdom, and she, speaking of God's creation, states:

A mist went up from the earth and watered the whole face of the ground. (Genesis 2:6)

> When He [God] prepared the heavens, I was there, when He drew a circle on the face of the deep, when He established the clouds above, when He strengthened the fountains of the deep... (Proverbs 8:27-28)

This passage seems to suggest that clouds were present in God's original creation even if rain was not.

By the way, Genesis 2:4-6 may provide an argument that deserts and ice sheets were absent in the pre-Fall and probably pre-Flood world. Note that the mist "watered the *whole* face of the ground" (emphasis added). If the mist was watering the ground everywhere, this would seem to rule out deserts and polar ice sheets.

A Pre-Flood Vapor Canopy?

For many years, respected creation scientists favored the idea of a water vapor canopy at the top of the earth's atmosphere. This vapor canopy was thought to help explain the different climate regime that existed in the pre-Flood world, as well as provide some of the water for the Flood.[21,22] I was a longtime fan of this particular hypothesis, but I have since concluded from scriptural considerations that this particular model isn't correct. Canopy advocates argued that this hypothetical vapor canopy corresponded to the "waters that were above the expanse" in Genesis 1:6-8 (ESV). However, the Bible explicitly states that God placed the stars in the expanse *between* the waters below and the waters above. If one assumes that the "waters above" are a vapor canopy above the earth's atmosphere, then the earth's atmosphere *is* the expanse. Clearly, no stars are present in the earth's atmosphere! It seems that expanse ("firmament" is actually a poor translation)[23] is all the space above the earth's surface, including interstellar space.[24,25] Hence, the "expanse" corresponds roughly to our word "sky."

This scriptural argument convinced me that the canopy model is incorrect. However, there are also potential scientific problems with the idea. Such a canopy might make the surface of the earth intolerably hot, as suggested by computer modeling experiments.[26] Regardless, the scientific arguments are equivocal because it is always possible in science that we have overlooked some details that might allow the model to work. But the argument from Scripture seems to me to decisively refute the canopy model.

Another Possible Explanation

Obviously, rain cannot occur without clouds. But clouds are composed of water droplets, not water in the gaseous state (water vapor). For droplets to form, water vapor must condense onto *cloud condensation nuclei* (CCNs). Cloud condensation nuclei are tiny particles, or droplets, called *aerosols* that have the necessary properties to serve as "seeds" for water droplets. Furthermore, even after the droplets form, rain will only occur if larger droplets are present. The smaller droplets collide with larger droplets, resulting in even larger droplets that are heavy enough to fall as rain. If CCNs were much less abundant in the pre-Flood world, then could this suppress droplet formation sufficiently to allow clouds but prevent rain? In that case, clouds might still be present, but the droplets would not become large enough to cause rain. A number of sources for aerosols—such as forest fires, wind-blown dust, volcanic aerosols, and products of human combustion[27]—could reasonably be expected to be either rare or absent in the pre-Flood world. Interestingly, Drs. Henry Morris and John Whitcomb suggested this possibility in their groundbreaking 1961 book *The Genesis Flood*.[28]

Some CCNs would allow clouds to form, but the cloud droplets would remain small enough to prevent rainfall, at least over the land. However, this is probably not the whole answer. A former professional with the National Weather Service pointed out to me a number of difficulties with it that I had not previously considered.[29] Right now, I see this as a possible working hypothesis.

A Biblical View of Human Population

The first chapters of Genesis also give us important insights into other issues raised by environmentalists, such as human population. Environmentalists insist there are too many people on the planet, and they advocate measures (including, tragically, abortion) to keep the human population in check. In fact, worldwide a billion unborn babies have been deliberately killed before birth within the last hundred years.[30] Surely this will be one of the things for which God judges this world before Christ returns.[31]

But a deliberate restriction of the human population is contrary to what God commanded Adam and Eve to do. God told them to be fruitful and fill the earth.

> Then God blessed them, and God said to them, "Be fruitful and multiply; fill the earth and subdue it; have dominion over the fish of the sea, over the birds of the air, and over every living thing that moves on the earth." (Genesis 1:28)

God never rescinded this command to be fruitful and multiply. Furthermore, God repeated this command to Noah and his family after the Flood (Genesis 9:1). We humans have never fulfilled this command. Much, much land is still available to be settled.

Despite the alarmist claims of many environmentalists, the earth is not overcrowded with people. One can easily verify this by looking out the window of an airplane as it crosses over the central United States. Yes, some cities are too crowded, but the obvious solution is to move out from those cities and build other cities, not to restrict the human population.

Are there really too many children?

> Enlarge the place of your tent, and let them stretch out the curtains of your dwellings; do not spare; lengthen your cords, and strengthen your stakes. For you shall expand to the right and to the left, and your descendants will inherit the nations, and make the desolate cities inhabited. (Isaiah 54:2-3)

As an example, nearly all China's population of over a billion people lives within the eastern half of the country. A very small fraction of the population occupies the remaining western half.[32] So, there is still plenty of land to go around. Granted, this western land in China may not be prime real estate, but it *is* usable.

Should Religious Beliefs Guide Environmental Policy?

Of course, environmentalists would argue that basing environmental views on "one's personal religious beliefs" is outrageously irresponsible. They would argue that encouraging, rather than discouraging, young couples to have more children is an invitation to an overpopulation disaster. But this is why I spent so much time discussing the evidence for the Genesis Flood. The water-deposited rocks, as well as the fossils within them, bear mute testimony that the Genesis Flood really happened. We no longer have direct access to the original pre-Fall and pre-Flood worlds, but we *do* have access to the rocks and fossils. Since the Bible's testimony about the Flood is true, then we should also take the rest of the Bible seriously, including those parts that deal with environmental issues.

Of course, conservative Christians can use this same logic to argue that environmental policies based on an evolutionary worldview are *themselves* dangerous because they are based on a wrong understanding of history. And this is exactly what I argue in the following chapters, especially in chapters 16-19.

Behind the claim that "you can't base public policy on your religious beliefs" is the unspoken assumption that those religious beliefs are wrong. The objection isn't to religious beliefs per se, because everyone has "religious beliefs," even atheists. Everyone has a worldview, so everyone has religious beliefs of some kind. The real issue is: Which religious beliefs are correct, and which are wrong? I am arguing that the religious beliefs of an evolutionary worldview are wrong, therefore they should not serve as the basis for environmental or climate policy. Instead, the biblical worldview is the correct one, and it should guide our thinking on these issues. One must take Genesis seriously in order to have a correct understanding of environmental issues confronting us today. The first chapters of Genesis are not incidental to thinking rightly about these issues—they are critical.

> "One must take Genesis seriously in order to have a correct understanding of environmental issues confronting us today."

References

1. Morris, H. M. III. 2016. *The Book of Beginnings: A Practical Guide to Understanding Genesis*. Dallas, TX: Institute for Creation Research, 277.
2. Thomas, B. 2017. Fast-Changing Killifish Swim Past Evolution. *Acts & Facts*. 46 (2): 15.
3. Thomas, B. 2014. Birds' Built-In Defenses Fend Off Radiation. *Creation Science Update*. Posted on ICR.org May 12, 2014, accessed November 1, 2018.
4. Thomas, B. Amazing Fish Adaptive Design. *Creation Science Update*. Posted on ICR.org May 18, 2012, accessed November 1, 2018.
5. Luyendyk, B. P. Oceanic crust. *Encyclopaedia Britannica online*. Posted on britannica.com, accessed June 3, 2014.
6. Cronin, T. M. 2009. *Paleoclimates: Understanding Climate Change Past and Present*. New York: Columbia University Press, 72.
7. Baumgardner, J. 2005. Carbon-14 Evidence for a Recent Global Flood and a Young Earth. In *Radioisotopes and the Age of the Earth: Results of a Young-Earth Creationist Research Initiative*, vol. 2. L. Vardiman, A. A. Snelling, and E. F. Chaffin, eds. El Cajon, CA and Chino Valley, AZ: Institute for Creation Research and the Creation Research Society, 587-630.
8. Clarey, T. Fossil Trees in Antarctica Preserve Ancient Proteins. *Creation Science Update*. Posted on ICR.org December 11, 2017, accessed May 8, 2018.
9. Hoesch, W. 2006. Arctic Heat Wave. *Acts & Facts*. 35 (8).
10. Carbon Dioxide through Geologic Time. 2002. Climate Change Past and Future: Climate and CO_2 in the Atmosphere. University of California, San Diego. Posted on earthguide.ucsd.edu, accessed December 12, 2018.
11. Cronin, *Paleoclimates: Understanding Climate Change Past and Present*, 69.
12. Yapp, C. J. and H. Poths, 1992. Ancient atmospheric CO_2 pressures inferred from natural goethites. *Nature*. 355: 342-344.
13. Humphreys, D. R. 2009. God's global warming worked just fine. Creation Ministries International. Posted on creation.com August 11, 2009, accessed May 8, 2018.
14. Ibid.
15. Hayhoe, K. Quoted in *Decoding the Weather Machine*. NOVA Public Broadcasting System documentary. Premiered April 18, 2018, transcript accessed October 8, 2018.
16. Lee, H. What does past climate change tell us about global warming? Skeptical Science. Posted on skepticalscience.com, accessed December 12, 2018. Emphasis in the original.
17. Ibid, emphasis in the original.
18. Sanford, J. 2014. *Genetic Entropy*. Waterloo, NY: FMS Publications.
19. Thomas, B. Did Adam Really Live 930 Years? *Creation Science Update*. Posted on ICR.org July 14, 2014, accessed December 11, 2018.
20. There is disagreement among creationists as to whether Peleg's name (which means "division") is referring to the division of peoples at the Tower of Babel or to geographic division by waterways and rivers that formed during the post-Flood Ice Age. Regardless of which position one takes on this issue, the dramatic decrease in human longevity that occurred in his generation may very well be evidence that Peleg lived shortly after the Tower of Babel.
21. Whitcomb, J. C. and H. M. Morris. 1961. *The Genesis Flood: The Biblical Record and Its Scientific Implications*, Phillipsburgh, NJ: Presbyterian and Reformed Publishing Company, 255-258.
22. Dillow, J. C. 1982. *The Waters Above: Earth's Pre-Flood Vapor Canopy*, rev. ed. Chicago, IL: Moody Press.

23. The word "firmament" in many old translations incorrectly suggests that the expanse is "firm" or "solid," a fact seized upon by biblical skeptics to argue that the Bible contains scientific error. However, this idea is not demanded by the original Hebrew word *raqîa'*. Dr. Jim Johnson of the Institute for Creation Research has concluded, based upon a detailed word study, that the basic idea is one of thinness or emptiness, *not* metallicity or hardness. Because space is indeed thin, empty, or tenuous, this is a very good description of space (including interstellar space).
24. Humphreys, D. R. 1994. A Biblical Basis for Creationist Cosmology. In *Proceedings of the Third International Conference on Creationism*. R. E. Walsh, ed. Pittsburgh, PA: Creation Science Fellowship, 255-266.
25. Thomas, B. 2016. What Were the 'Waters Above the Firmament'? *Acts & Facts*. 45 (5): 20.
26. Vardiman, L. 2003. Temperature Profiles for an Optimized Water Vapor Canopy. In *Proceedings of the Fifth International Conference on Creationism*. R. Ivey, ed. Pittsburgh, PA: Creation Science Fellowship, 29-39.
27. Rogers, R. R. and M. K. Yau. 1989. *A Short Course in Cloud Physics*, 3rd ed. New York: Butterworth-Heinemann, 90.
28. Whitcomb and Morris, *The Genesis Flood*, 241.
29. Personal correspondence with David Vonderheide, a former hydrometeorological technician at the National Weather Service.
30. Jacobson, T. and W. R. Johnston. 2019. *Abortion Worldwide Report: 1 Century, 100 Nations, 1 Billion Babies.* West Chester, OH: GLC Publications.
31. Even those not necessarily convinced that life begins at conception should tremble at what our civilization has done. Abortion truly "stops a beating heart." It is obvious from Scripture that God expects society to protect the life of the unborn child (Exodus 21:22-25, especially verse 23), and any society that fears God must err on the side of caution regarding this issue.
32. Wood, P. Population Density in China. Commentary on East Asian Security Issues & Foreign Policy. Posted on p-wood.co May 16, 2017, accessed December 11, 2018.

7 Past Climate Change in a Uniformitarian Worldview

Summary: Secular scientists are unable to convincingly explain past climate change. Furthermore, their belief that the sun is billions of years old leads to a well-known problem called the young faint sun paradox. Secular scientists think that billions of years ago the sun would have been 30% dimmer than it is today, giving that much less energy to Earth's climate. This would have caused Earth to be a frozen wasteland, despite the fact that most evolutionists think Earth was quite warm at that time. Secular scientists believe there were five major Ice Ages in Earth's supposed 4.6-billion-year history, with 50 smaller Ice Ages (or glacial intervals) occurring during the last major Ice Age. Evidence for a single, recent Ice Age is strong, but evidence for multiple Ice Ages is very weak. And for reasons explained in later chapters, secular scientists don't have an adequate explanation for what caused these supposed multiple Ice Ages.

Uniformitarian scientists insist that Earth is 4.6 billion years old and that it formed, along with the rest of the solar system, from a rotating cloud of dust and gas. Likewise, most claim that the first single-cell organism spontaneously arose about 3.8 billion years ago.[1]

Interestingly, this multi-billion-year timescale itself results in a challenge to the evolutionary story known as the *young faint sun paradox*. Based on their models of how stars slowly change over time, uniformitarians believe that about 4 billion years ago the

Secular scientists claim that our solar system formed billions of years ago from a rotating cloud of gas and dust

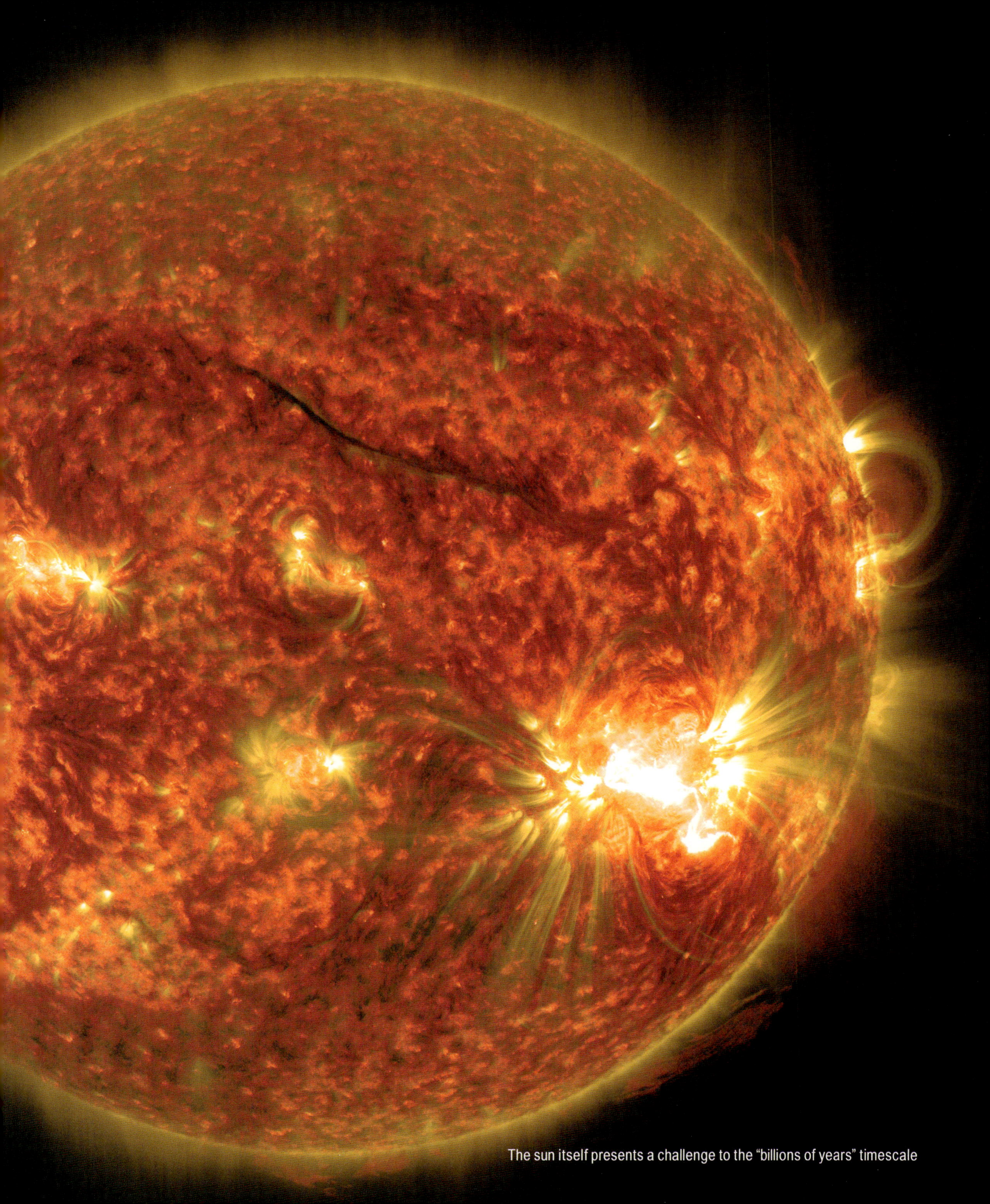

The sun itself presents a challenge to the “billions of years” timescale

solar energy output would have been only 70% of what it is today.[2,3] This is because the process of nuclear fusion would gradually change the composition and size of the sun. As the aging sun decreased in size, it would become easier for the process of nuclear fusion to proceed, causing the sun to become brighter.[4] This, of course, implies that the sun was dimmer in the past. In the absence of other factors to compensate for this decreased energy output (such as a greater concentration of greenhouse gases in the atmosphere), the earth would have been much colder around 4 billion years ago, cold enough for the oceans to freeze over. This is a problem for evolutionists because most of them think Earth's climate was actually quite warm at that time, with liquid water present on Earth when the first organisms supposedly evolved.[5]

> "This multi-billion-year timescale itself results in a challenge to the evolutionary story known as the *young faint sun paradox.*"

Uniformitarian scientists have been aware of this problem for decades. Occasionally, the popular media claims that this problem has been solved—but these claims are short-lived.[6] Even as recently as 2016, the issue was still unresolved, although secular scientists claimed to be close to finding a solution.[7] Of course, the young faint sun paradox is nonexistent if the sun is only 6,000 years old.[8] Only over timescales of billions of years would one expect big changes in solar energy output due to a changing size and composition of the sun. No such drastic change in energy output should occur if the sun is only 6,000 years old.

Uniformitarian scientists claim there have been at least four major Ice Ages in Earth's very ancient past. The first is thought to have occurred billions of years ago, and the other three supposedly occurred hundreds of millions of years ago. Some secular scientists believe that there were times during the second Ice Age that the entire

Some secular scientists think the earth was completely frozen in the distant past

Glacial striations in a rock from British Columbia, Canada

earth was covered in ice, an event they call "Snowball Earth."[9,10]

However, the evidence for these supposed ancient Ice Ages is very weak. One of the reasons that uniformitarians believe the Snowball Earth scenario is because they have found scratches in rocks they think were close to the equator hundreds of millions of years ago. Such scratches in rocks can be caused by glaciers. Rocks embedded in glacial ice often scratch the underlying bedrock as the glacier advances. So, some uniformitarians have concluded that glaciers were large enough to stretch from the polar regions all the way to the equator, which would imply that the entire earth was covered in ice.

However, there is a better explanation for these scratches. They can also be caused by debris flows such as avalanches or landslides. Creation scientists have made strong arguments that the scratches being used as evidence for past Ice Ages actually resulted from giant underwater landslides (Figure 7.1)—the kind that one would expect in a cataclysmic, global Flood.[11-13] In fact, it was a uniformitarian geologist who pointed out that his fellow uniformitarians were likely incorrectly attributing these scratches to glaciers.[14,15] Moreover, secular scientists don't have good explanations for why these

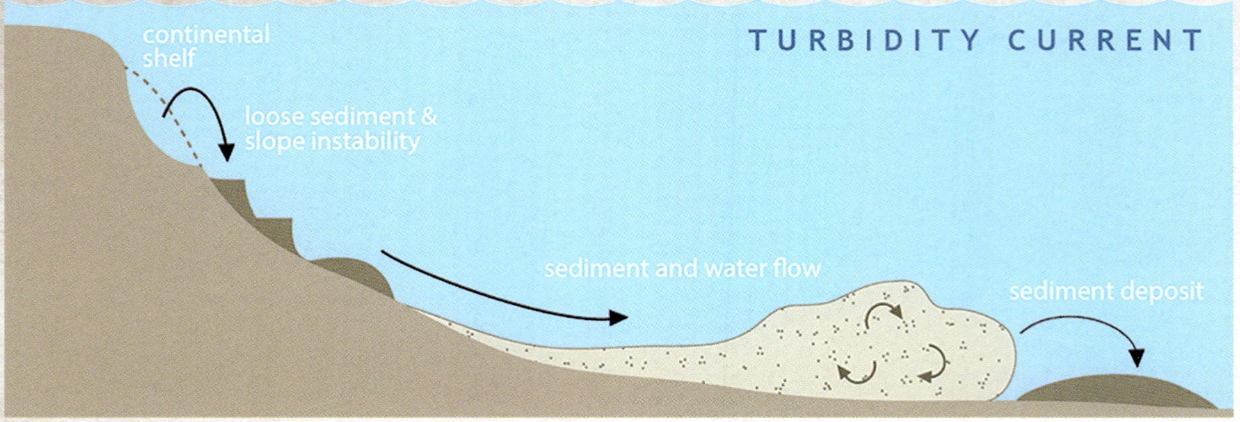

Figure 7.1. Underwater debris flows such as turbidity currents can cause scratches in rocks similar to those formed by glaciers

supposed ancient Ice Ages occurred or why they ended. Their explanations are very speculative.[16,17]

Finally, secular scientists believe that a fifth major Ice Age began 2.6 million years ago at the beginning of what they call the Pleistocene epoch. This most recent major Ice Age is called the Quaternary glaciation. They claim that within this major Ice Age, there were 50 or so glacial cycles during which the ice sheets grew and advanced.[18] These glacial cycles are what most people refer to as "Ice Ages" in popular speech. Supposedly, between the glacials were warmer periods called *interglacials*. Within the last million years or so, these glacial–interglacial cycles are said to have been about 100,000 years long. The glacial parts of the cycles lasted around 80,000 years, and the warmer interglacials lasted about 20,000 years. Prior to that time, the Ice Age cycles were thought to last about 41,000 years.

"In fact, it was a uniformitarian geologist who pointed out that his fellow uniformitarians were likely incorrectly attributing these scratches to glaciers."

That, in a nutshell, is what uniformitarian scientists believe about past climate change. Although there is strong geological evidence for a single recent Ice Age, the supposed evidence for these other Ice Ages, including multiple Pleistocene Ice Ages, is much weaker. Furthermore,

Perito Moreno Glacier in Patagonia, Argentina

uniformitarian explanations are inadequate to explain this Ice Age.

References

1. Bell, E. A. et al. 2015. Potentially biogenic carbon preserved in a 4.1 billion-year-old zircon. *Proceedings of the National Academy of Sciences.* 112 (47): 14518-14521.
2. Cronin, T. M. 2010. *Paleoclimates: Understanding Climate Change Past and Present*. New York: Columbia University Press, 58.
3. Feulner, G. 2012. The Faint Young Sun Problem. *Reviews of Geophysics.* 50 (2): RG2006.
4. Sarfati, J. 2004. Our steady sun: a problem for billions of years. *Creation.* 26 (3): 52-53.
5. Cronin, *Paleoclimates*, 59.
6. Netburn, D. Mystery of the 'faint young sun paradox' may be solved. *Los Angeles Times.* Posted on latimes.com July 10, 2013, accessed January 19, 2017.
7. Rathi, A. A new theory is close to solving one of the greatest mysteries of how life began on Earth. *Quartz.* Posted on qz.com May 25, 2016, accessed January 19, 2017.
8. Faulkner, D. 2001. The young faint Sun paradox and the age of the solar system. *Journal of Creation.* 15 (2): 3-4.
9. Cronin, *Paleoclimates*, 59-64.
10. Ice Age. *History.com.* Posted on history.com March 11, 2015, accessed November 14, 2016.
11. Molén, M. 1990. Diamictites: Ice-Ages or Gravity Flows? In *Proceedings of the Second International Conference on Creationism.* R. E. Walsh and C. L. Brooks, eds. Pittsburgh, PA: Creation Science Fellowship, 177-190.
12. Oard, M. J. 1997. *Ancient Ice Ages or Gigantic Submarine Landslides?* Chino Valley, AZ: Creation Research Society Books.
13. Oard, M. J. 2009. Landslides Win in a Landslide over Ancient "Ice Ages." In *Rock Solid Answers: The Biblical Truth Behind 14 Geological Questions.* M. Oard and J. K. Reed, eds. Green Forest, AR: Master Books.
14. Schermerhorn, L. J. G. 1974. Late Precambrian Mixtites: Glacial and/or Nonglacial? *American Journal of Science.* 274: 673-824.
15. Schermerhorn, L. J. G. 1974. No evidence for glacial origin of late Precambrian tilloids in Angola. *Nature.* 252: 114-115.
16. Marshall, M. The history of ice on Earth. *New Scientist.* Posted on newscientist.com May 24, 2010, accessed November 27, 2017.
17. Maasch, K. A. What Triggers Ice Ages? PBS. Posted on pbs.org December 31, 1996, accessed February 27, 2017.
18. Walker, M. and J. Lowe. 2007. Quaternary Science 2007: a 50-year Retrospective. *Journal of the Geological Society, London.* 164 (6): 1073-1092.

8 Uniformitarian Ice Age Explanations Are Inadequate

Summary: Secular scientists consider the cause of Ice Ages one of the "great mysteries of science." Cool summers and moderate winters for many consecutive years are necessary to create an Ice Age, but secular scientists don't have an adequate mechanism for this. The Milankovitch theory is the most popular secular explanation. It says that Earth's tilt wobbles as it orbits the sun, changing how much sunlight hits the earth in different places. These wobbles would have resulted in less summer sunlight in the northern polar regions at different times in the prehistoric past. This reduced summer sunlight presumably would have allowed the ice sheets to grow, resulting in intermittent Ice Ages lasting thousands of years. However, multiple things about this theory are puzzling, nine of which are briefly discussed in this chapter. Moreover, if Earth's climate is stable, this mechanism isn't strong enough by itself to cause Ice Ages. Therefore, secularists assume that Earth's climate is unstable. Thus, much of climate change alarmism can be traced back to the inadequacies of the Milankovitch theory.

Since uniformitarians claim that Ice Ages played a major part in Earth history, you might think they have a convincing explanation for what causes Ice Ages. Actually, they don't. By the late 1960s, secular scientists had proposed over 60 different Ice Age theories. The number is likely much higher today.[1] In 1997, *U.S. News & World Report* stated that the cause of an Ice Age or Ages was one of the "great mysteries of science."[2]

> Scientists have long known about the giant ice sheets that the Cenozoic ushered in. Even in the mid-19th century they knew that

> glaciers had repeatedly raked swaths of Europe and North America in the not so distant past. Yet despite the efforts of marine geologists, atmospheric chemists, oceanographers, and more, no one knows what caused the ice ages. "We've been chewing on this problem for 30 or 40 years," says Alan Mix, an oceanographer at Oregon State University. "It's a killer." Adds Ralph Cicerone, dean of physical sciences at the University of California–Irvine, "It's embarrassing."[2]

Not much has changed since 1997. Scientists are still proposing new theories,[3,4] an indication that the other theories don't work very well.[5] The precise causes for an Ice Age or Ages "remain controversial"[6] and what causes Ice Ages "is not completely understood."[7] One writer noted, "Although theories abound, no one really knows what causes Ice Ages."[8] Secular scientists make conflicting claims. Most claim that an astronomical influence of some kind is affecting Ice Ages (more on that later), but others disagree.

"Although theories abound, no one really knows what causes Ice Ages."

Why Is an Ice Age So Difficult to Explain?

An Ice Age requires two things. First, summers must be cool enough to prevent winter snow and ice from completely melting. If snow and ice don't melt during the summer, then more snow and ice will accumulate the following winter. If such conditions persist for many years, then thick ice sheets will form over time.

Another requirement is heavy snowfall. This is necessary because even with cooler summers, snow and ice will still melt. Solar radiation can melt the snow, even when temperatures are below freezing. You might think that colder winters are needed for heavier snowfall, but this isn't the case. Moderate, or even warmer winters, are required. This is because there is much less moisture in colder air than in warmer air.[9] Less moisture in winter air means less snowfall, not more.

So, cooler summers and heavy snowfall are needed to

cause an Ice Age. This means that cool winters and moderate (or even warmer) winters are required. However, it is difficult to see how both requirements can occur at the same time, especially for thousands of years. To prevent snow and ice from melting, you need colder summers. Furthermore, since secular scientists think Ice Ages last thousands of years, these colder summers would also have to persist for thousands of years. But if summer temperatures were colder for thousands of years, then it seems obvious that the winters would also have to be colder. After all, we tend to associate colder summers with a generally colder climate. You don't expect temperatures to go back and forth for tens of thousands of years between lower-than-average summer temperatures and higher-than-average winter temperatures. Rather, over a long time, we expect cold winters generally to accompany cold summers or warm winters to accompany warm summers. An Ice Age, especially one lasting thousands of years, seems to require very abnormal conditions, conditions that seem very unlikely in today's world. This is the basic problem with uniformitarian Ice Age theories.

The Milankovitch Ice Age Theory

However, one particular Ice Age theory is very popular among uniformitarian scientists. It is called the *astronomical Ice Age theory* because it involves Earth's motions in space as it travels around the sun. It is also called the *Milankovitch* (pronounced mih-LAWN-koe-vitch) *theory* in honor of Serbian scientist Milutin Milanković, who was a prominent advocate for the theory.[10] Because the Milankovitch theory is currently the most popular, I will spend a great deal of time discussing it.

As most of us learned in school, the earth's axis has a tilt of about 23.5°. This means the axis makes an angle of 23.5° with a line perpendicular to the plane of the earth's orbit around the sun. Gravitational influences from other solar system bodies cause this tilt to slowly change. Given enough time, the tilt (also called the *obliquity*) would vary between 22° and 24.5°. Astronomical calculations show that it would take 41,000 years for the tilt to start at a minimum value of 22°, increase to the maximum value of 24.5°, and then decrease back to the minimum value of 22° (Figure 8.1).

Milutin Milanković

Of course, the 6,000 years of biblical history is not enough time for even one full tilt cycle. But because uniformitarian scientists believe the age of the solar system is truly vast, they feel free to run the numbers and extrapolate these tilt changes into both the distant past and distant future. For this reason, they think they know what the earth's tilt was millions of years ago.

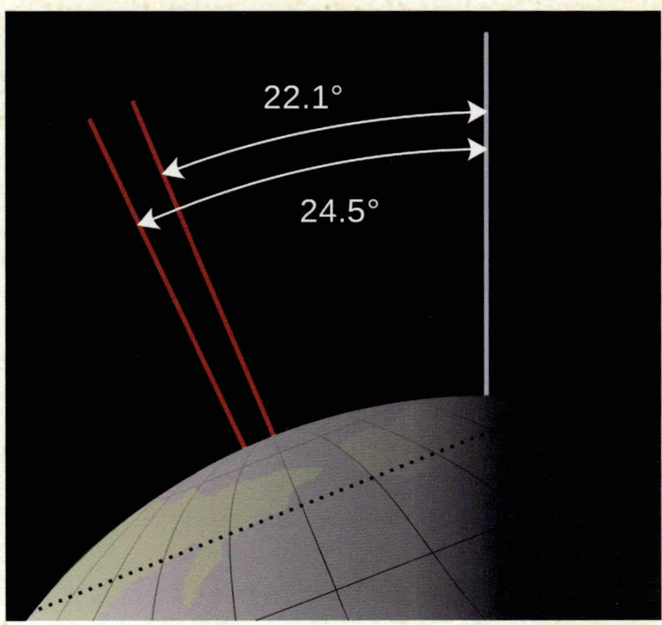

Figure 8.1. In a billions-of-years-old solar system, the tilt of the earth's axis would exhibit a 41,000-year cycle

As the earth orbits the sun, it traces out a path called an *ellipse*, which looks like a squashed circle. Given enough time, the shape of the earth's orbit would also undergo cycles. There would be times when the orbit was slightly more circular, and times when it was slightly less circular (Figure 8.2). A quantity called *eccentricity* indicates the shape of an orbit. The eccentricity can be between 0 and 1. An eccentricity of zero indicates a perfect circle, and an eccentricity approaching 1 indicates an extremely stretched or elongated ellipse. In a supposedly ancient solar system, Earth's elliptical orbit will exhibit cycles having lengths of 95K, 125K, and 405K years. However, the first two of these cycles are often lumped together and treated as a single cycle of about 100,000 years. In addition to its changing shape, the earth's orbital ellipse is slowly rotating relative to the background stars.

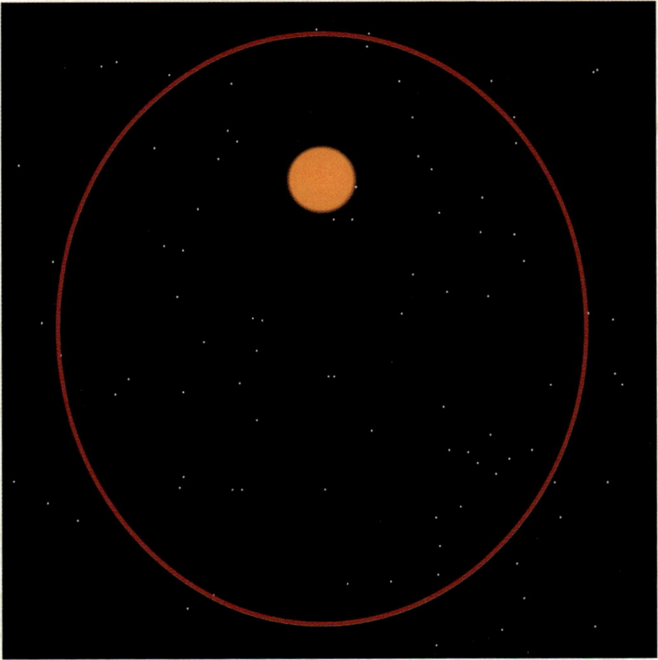

Figure 8.2. Over time, the earth's axis becomes more and then less elliptical

Likewise, the earth's rotational axis will also wobble like a top, given enough time. The length (or period) of this *precessional* wobble is 26,000 years (Figure 8.3). However, this wobble combines with the rotation of the earth's orbit to produce an overall *precessional cycle* that is 23,000 years long. At certain times, a second precessional cycle of 19,000 years would also be present.

Creation scientists don't disagree with the physics and astronomy used to perform these calculations. But nearly all these calculations are meaningless, since the solar system is just 6,000 years old. It is also important to realize that these calculations in no way prove a vast age of the solar system. When secular scientists do these calculations, they are simply assuming that the solar system is very, very old, so they feel free to perform these calculations for times in the prehistoric past.

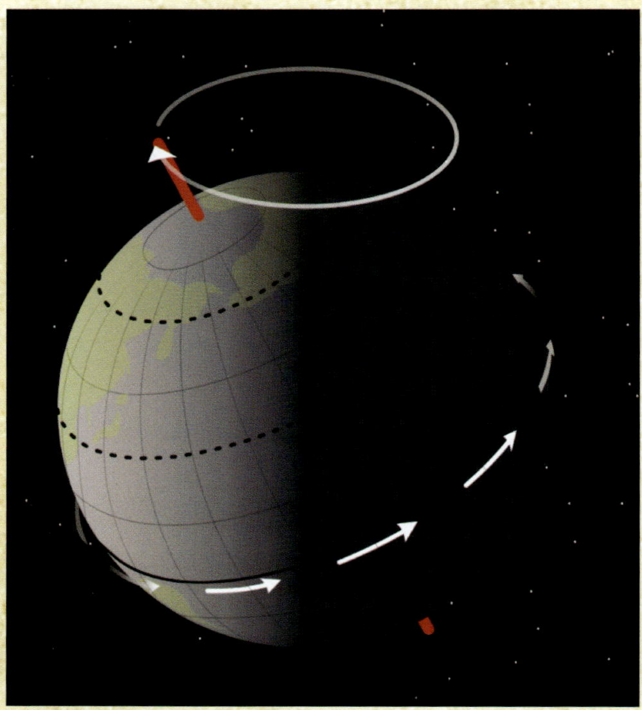

Figure 8.3. The earth wobbles like a top so that its spin axis exhibits a 26,000-year precessional cycle

As a result of these changes in Earth's orbital and rotational motions, the way sunlight strikes the earth would vary over many thousands of years. At different times, some latitudes would receive more or less sunlight than other latitudes. Likewise, at different times some seasons would receive more or less sunlight than other seasons. Most uniformitarian scientists think these variations in sunlight controlled, or paced, the timing of the 50 Pleistocene Ice Ages that supposedly occurred in the last 2.6 million years. They think that when less summer sunlight falls on the northern polar ice sheets, this reduced sunlight allows the ice sheets to grow in size. The result is a glacial interval, or Ice Age. Likewise, when more summer sunlight falls on these northern ice sheets, the increased sunlight melts the ice sheets, causing them to get smaller in size. The result is a warm interglacial.

Problems with the Milankovitch Theory

Despite the popularity of this theory, it has serious problems. In fact, it really ought to be called a hypothesis rather than a theory due to these problems and a lack of strong supporting scientific evidence (more on that later). Its biggest problem is that these changes in sunlight distribution are very small. So, it is hard to see how those alone could be the cause of an Ice Age or Ages. Secular astronomer Fred Hoyle was a skeptic of the Milankovitch theory. He once made the following quip to illustrate the difficulty of attributing past Ice Ages to subtle changes in sunlight:

> If I were to assert that a glacial condition could be induced in a room liberally supplied during winter with charged night-storage heaters simply by taking an ice cube into the room, the proposition would be no more unlikely than the Milankovitch theory.[11]

Hoyle's skepticism was well-justified. In fact, the Milankovitch theory suffers from many well-known mysteries and paradoxes. In other words, there are things about the theory that don't make sense. These issues are acknowledged even by secular scientists.[12,13]

Milankovitch Mysteries

What are some of these problems? The following is summary of some of the major mysteries about the theory.

1. *The General Weakness of the Mechanism*. As noted earlier, the changes in sunlight distribution are so small that it's difficult to see how they alone could be responsible for a changing climate.

According to the Milankovitch (or astronomical) theory, changes in Earth's rotational and orbital motions will subtly change the way that sunlight is distributed with season and latitude over many thousands of years, and these changes are thought to control the timing of Ice Ages.

2. *The 100,000-Year Problem.* Uniformitarian scientists believe that within the last million years, Ice Age cycles were roughly 100,000 years long. At first glance, this seems to be an argument for the theory since one of the dominant eccentricity cycles is about 100,000 years long. That seems to make sense, right? However, the variations in sunlight caused by this eccentricity cycle are the smallest of all the variations resulting from Earth's orbital motions.[14] All the variations in sunlight distribution are small, but these are the smallest. Why would the climate be responding to such tiny changes?
3. *The Unsplit 100,000-Year Peak Problem.* Close examination reveals that the 100,000-year cycle is actually made of two cycles: a 95,000-year cycle and a 125,000-year cycle. Yet, most climate data sets do not show evidence for these two cycles. For this reason, some uniformitarian scientists have proposed an alternate explanation for the 100,000-year cycle.[15]
4. *Mid-Pleistocene Transition Problem.* Uniformitarians claim that about a million years ago, in the middle of the Pleistocene, the length of climate cycles changed from 41,000 years to 100,000 years. However, there is no obvious astronomical reason for this change. Recently, some uniformitarian scientists claimed to have solved this problem, but it remains to be seen whether this claim will stand up to scrutiny.[16]
5. *The Causality Problem.* One of the more embarrassing problems for the Milankovitch theory is that if uniformitarian interpretations of some data sets are to be believed, then the earth started to come out of the second-most-recent (penultimate) Ice Age thousands of years before the increases in sunlight that supposedly caused this warming.[17] In other words, the effect (a warmer climate) seems to be coming *before* its cause (increases in sunlight)!

6. *The Stage 11 Problem*. A roughly 50,000-year interval centered about 400,000 years ago is known as Marine Isotope Stage 11. (The concept of marine isotope stages is explained later.) This particular time interval is a problem because it was supposedly a very long and warm interglacial period. Yet, changes in sunlight distribution at that time cannot account for it. For this reason, uniformitarian computer simulations have been unable to successfully model it.[18]
7. *The Phase Problem*. When the northern high latitudes receive less summer sunlight, the high southern latitudes receive more sunlight. So, one would expect Ice Ages to occur at different times in different hemispheres. In other words, one expects northern and southern hemisphere Ice Ages to be "out of phase" with each other. Yet, uniformitarian scientists think that large ice sheets have been "in phase," present in both hemispheres at the same times.[19,20]

8. *The Pliocene Transition Problem.* According to the Milankovitch theory, Ice Age cycles started about 2.6 million years ago, at the end of the Pliocene epoch and the beginning of the Pleistocene epoch. But since the earth was presumably undergoing orbital and rotational motions for *billions* of years before this, then why did Ice Age cycles start only 2.6 million years ago? What was so different about the most recent 2.6 million years that allowed Ice Ages to occur when Ice Ages were not occurring before this? Some might argue that plate tectonics played a role and that the earlier positions of the continents somehow prevented Ice Ages from occurring. However, the continents were, by secular reckoning, already in their present locations millions of years *before* the Pleistocene began. Moreover, other uniformitarian scientists do not consider this explanation satisfactory since they have proposed other hypotheses to account for this.[21]

9. *The 41,000-Year Paradox.* Uniformitarians believe that between 2.6 and 1 million years ago, Ice Age cycles were 41,000 years long. However,

changes in sunlight at this time due to the 23,000-year precessional cycle should have been larger than those due to the 41,000-year cycle. Why then did the climate respond to the weaker 41,000 cycle?[22]

Given all these problems, one might wonder why the Milankovitch theory is so popular. We will address that question in the next chapter.

References

1. Oard, M. J. 1990. *Ice Age Caused by the Genesis Flood.* El Cajon, CA: Institute for Creation Research, 13.
2. Watson, T. 1997. What causes ice ages? *U.S. News & World Report.* 123 (7): 58-60.
3. Nie, J. et al. 2014. Pacific Freshening Drives Pliocene Cooling and Asian Monsoon Intensification. *Scientific Reports.* 4 (5474).
4. New theory on cause of ice age 2.6 million years ago. *ScienceDaily.* Posted on sciencedaily.com June 27, 2014, accessed December 12, 2016.
5. Oard, M. 2006. *Frozen in Time: Woolly Mammoths, the Ice Age, and the Biblical Key to Their Secrets.* Green Forest, AR: Master Books, 64.
6. Tanabe, J. Ice age. *New World Encyclopedia.* Posted on newworldencyclopedia.org October 12, 2016, accessed May 7, 2019.
7. Ice Ages. BBC. Posted on bbc.co.uk, revised October 12, 2016, accessed December 12, 2016.
8. Alt, D. 2001. *Glacial Lake Missoula and Its Humongous Floods.* Missoula, MT: Mountain Press Publishing Company, 180.
9. Lutgens, F. K. and E. J. Tarbuck. 2010. *The Atmosphere: An Introduction to Meteorology,* 11th ed. New York: Prentice Hall, 102-103.
10. Milanković, M. 1941. *Canon of Insolation and the Ice-Age Problem.* Belgrade, Serbia: Special Publication of the Royal Serbian Academy, vol. 132.
11. Hoyle, F. 1981. *Ice: The Ultimate Human Catastrophe.* New York: Continuum, 77.
12. Cronin, T. M. 2009. *Paleoclimates: Understanding Climate Change Past and Present.* New York: Columbia University Press, 130-139.
13. Oard, M. J. 2007. Astronomical troubles for the astronomical hypothesis of ice ages. *Journal of Creation.* 21 (3): 19-23.
14. Cronin, *Paleoclimates,* 130.
15. Muller, R. A. and G. J. MacDonald. 2002. *Ice Ages and Astronomical Causes: Data, Spectral Analysis and Mechanisms.* Chichester, UK: Praxis Publishing, 13, 40-45.
16. Lee, J.-E. et al. 2017. Hemispheric sea ice distribution sets the glacial tempo. *Geophysical Research Letters.* 44 (2): 1008-1014.
17. Winograd, I. J. et al. 1992. Continuous 500,000-Year Climate Record from Vein Calcite in Devils Hole, Nevada. *Science.* 258 (5080): 255-260.
18. Droxler, A. W. et al. 2003. Unique and Exceptionally Long Interglacial Marine Isotope Stage 11: Window into Earth Warm Future Climate. In *Earth's Climate and Orbital Eccentricity: The Marine Isotope Stage 11 Question.* AGU Monograph 137. Washington, DC: American Geophysical Union, 1-14.
19. Weber, M. E. et al. 2011. Interhemispheric ice-sheet synchronicity during the last glacial maximum. *Science.* 334 (6060): 1267.
20. Oard, M. J. 2014. Phase problems with the astronomical theory. *Journal of Creation.* 28 (2): 11-13.
21. Paillard, D. 2006. What Drives the Ice Age Cycle? *Science.* 28 (313): 455-456.
22. Ibid, 455.

9 The Climate Scandal You Haven't Heard About

Summary: The field of climate studies has seen its share of controversies in recent years, including a number of scandals that call into question the objectivity of some climate researchers. These include the controversy over Penn State climatologist Michael Mann's famous "hockey stick" graph, which purported to show extreme warming at the end of the 20th century. Although this graph became an icon of global warming activism, it is now widely accepted that the graph was seriously flawed, if not flat-out fraudulent. In 2009, a hacker leaked emails that showed climate researchers were using hardball tactics to squelch debate on global warming. However, there is another climate scandal of sorts that has gone unnoticed, even by many who are well-informed on the global warming controversy. This scandal involves evidence for the Milankovitch theory, used by secular scientists to explain past Ice Ages. The modern version of the theory originated from a famous "Pacemaker of the Ice Ages" paper published in 1976. Evidence for the theory depended critically upon an assumed age for the most recent flip, or reversal, of the earth's magnetic field. However, that assumed age was revised decades later by Milankovitch proponents themselves. This seriously undermined evidence for the Milankovitch theory—though hardly anyone seems to have noticed!

Climatology, the study of climates, has seen its share of controversy. One such controversy was Penn State University climatologist Michael Mann's famous (or infamous) "hockey stick" temperature graph. His graph seemed to show that northern hemisphere temperatures at the end of the 20th century were the highest they had been in 1,000 years.[1] This graph quickly become a powerful symbol of global warming. However, Mann's analysis has been widely criticized, even by many who think man-made global warming is real.[2,3]

Additionally, in 2009 a hacker leaked emails from the Climate Research Unit at

Great Britain's East Anglia University. The leaked emails quickly created a firestorm of controversy because they seemed to show that prominent climate scientists were using hardball tactics to squelch debate on the global warming issue.[4]

> "The leaked emails quickly created a firestorm of controversy because they seemed to show that prominent climate scientists were using hardball tactics to squelch debate on the global warming issue."

These two are by no means the only climate controversies or scandals, but they are among the most famous. However, another climate scandal of sorts has long remained under the radar, one that is possibly even more important than these other controversies. In the last chapter, we discussed some of the problems confronting the Milankovitch Ice Age theory. In spite of the theory's many problems, it is now widely accepted by secular scientists. This is largely because of a famous 1976 paper in the journal *Science* titled "Variations in the Earth's Orbit: Pacemaker of the Ice Ages."[5]

The Pacemaker of the Ice Ages

Research ships such as the *JOIDES Resolution* can drill into the sediments on the ocean floor. Scientists then extract the cores they have drilled (Figure 9.1). They study data within those sediments in an attempt to infer past changes in climate.

Oceanographers often use research vessels such as the *JOIDES Resolution* to drill and extract deep-sea sediment cores from the ocean floor

Figure 9.1. Close-up of a deep-sea sediment core

The authors of the Pacemaker paper used uniformitarian assumptions to analyze chemical "wiggles" in two deep-sea sediment cores from the southern Indian Ocean. They used the data from these cores to argue that astronomical cycles were influencing climate. In doing so, they also made indirect use of data from a third core in the far western Pacific Ocean (Figure 9.2). The chemical wiggles within the two Indian Ocean cores seemed to show climate cycles having lengths of about 100K, 42K, and 23K years. These cycle lengths were very close to the lengths of 100K, 41K, and 23K-year cycles that astronomers had already calculated for Earth's orbital and rotational motions. Secular scientists saw the close agreement between these numbers as strong evidence that the Milankovitch theory is correct.

The Pacemaker paper is crucial to uniformitarian scientists. It is the main reason the Milankovitch theory is widely accepted among secular scientists.

> The widely accepted Croll-Milankovitch theory that fluctuating climate conditions during the Quaternary glaciation have been driven by astro-

nomical cycles is based *entirely* on time-series analysis of paleoclimatic and orbital data.[6]

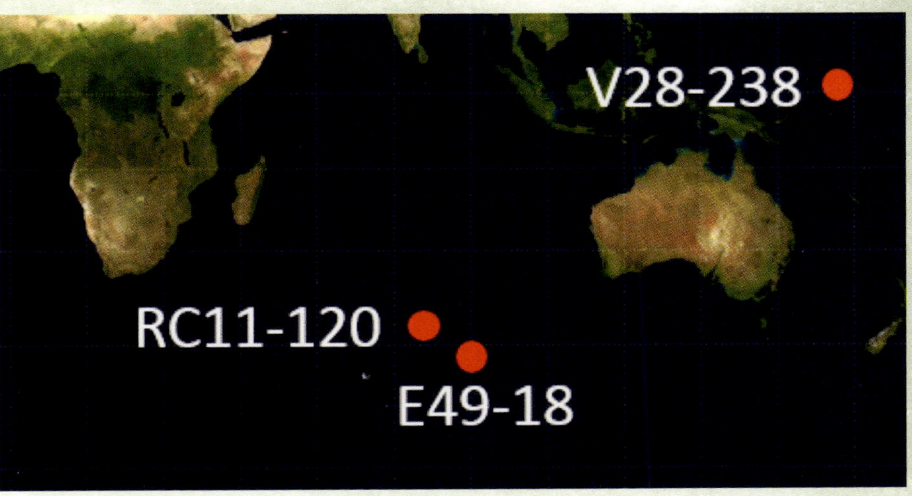

Figure 9.2. The famous "Pacemaker of the Ice Ages" paper used data from three deep-sea sediment cores, designated as RC11-120, E49-18, and V28-238

In fact, the evidence for the role of astronomy [in climate variation] comes almost exclusively from spectral analysis. The seminal paper was published in 1976, titled, "Variations of [*sic*] the earth's orbit: pacemaker of the ice ages."[7]

This paper is so important that the prestigious journals *Nature* and *Science* both ran articles commemorating its 40th anniversary.[8,9] However, original ICR research has revealed very serious problems with this paper, problems that are serious enough to invalidate its conclusions.[10-15]

In order to understand this paper's problems, it is necessary to discuss the measurements that scientists make within the deep-sea cores. This will also become important later as we discuss how the Milankovitch theory is influencing climate change speculation. The rest of this chapter has a lot of detail in it. In order for you to understand the scandal surrounding the Pacemaker paper, you need to know some basic terminology used by secular scientists. I have tried very hard to explain these concepts in simple terms (with no equations), but you may still have to read this chapter slowly and carefully. However, if you will hang in there, I think you'll agree that the payoff was definitely worth it. The bottom line is that an iconic, well-known argument for the Milankovitch theory is seriously flawed. Fortunately, with the possible exception of the appendixes, the reading in this chapter is as hard as it gets.

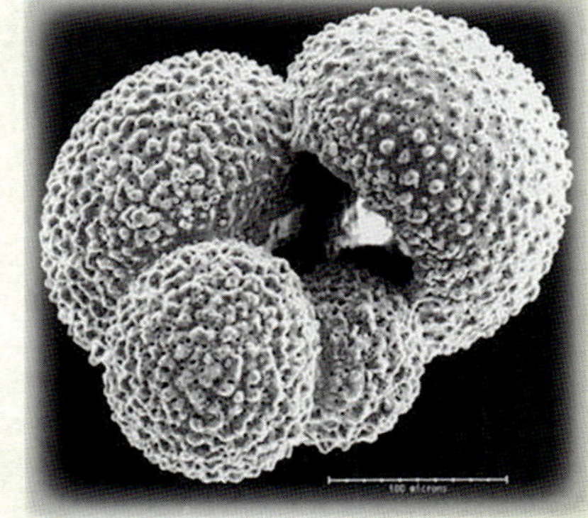

Figure 9.3. The authors of the "Pacemaker of the Ice Ages" paper used chemical measurements from the shells of the free-floating species of foraminifera known as *Globigerina bulloides*

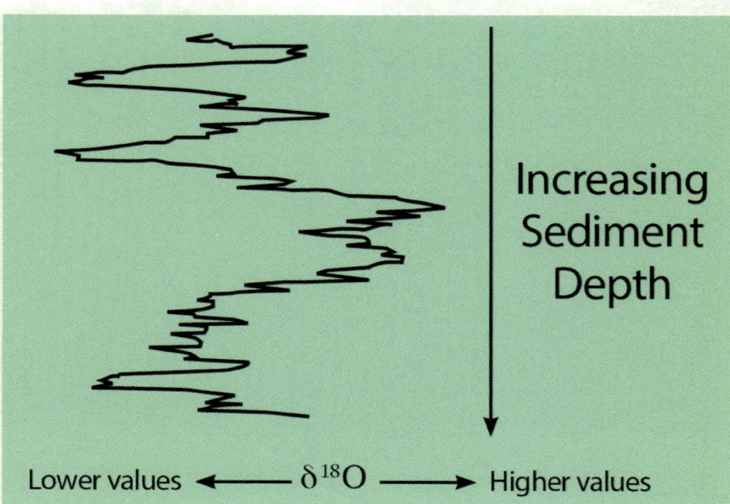

Figure 9.4. When oxygen isotope ratios from foraminifera shells in the seafloor sediments are plotted as a function of depth, many "wiggly" patterns can be seen. Secular scientists think these wiggles are telling a story about past climate change.

Oxygen Isotope Ratios

When microscopic marine creatures called *foraminifera* die, their shells become part of the debris accumulating on the ocean floor. These shells are made of calcium carbonate, $CaCO_3$, which contains oxygen (Figure 9.3). There are three stable varieties, or isotopes, of the oxygen atom: oxygen-16, oxygen-17, and oxygen-18. The chemical shorthand symbols for these three varieties of oxygen are ^{16}O, ^{17}O, and ^{18}O, respectively. Oxygen-16 is by far the most abundant isotope of oxygen. Oxygen-18 is roughly 500 times less abundant than oxygen-16, and oxygen-17 is extremely rare.

When foraminifera (also called forams) make their shells, they can use all three oxygen isotopes. Scientists can measure the amounts of ^{18}O and ^{16}O in a foram shell. They use these measurements to calculate a number called the *oxygen isotope ratio*. This number is indicated by the shorthand symbol $\delta^{18}O$. When scientists make graphs showing the different $\delta^{18}O$ values for different depths, many "wiggles" appear on the graph (Figure 9.4). Uniformitarian scientists believe these $\delta^{18}O$ wiggles are telling a story about climate change. They think $\delta^{18}O$ values communicate the total amount of land ice at the time the foram made its shell. High $\delta^{18}O$ values are thought to indicate the deeper parts of Ice Ages with *more* ice, and lower $\delta^{18}O$ values are thought to indicate times of *less* ice during the warmer periods, called *interglacials*.[16]

Wiggle Matching

Secular scientists think the $\delta^{18}O$ values are a global climate indicator. They believe that ideally the same basic oxygen isotope signal, or pattern, should be present in *all* seafloor sediment cores. They expect this to be true even in cores located thousands of miles apart. They also expect it to be true even if these similar-looking features are located at different depths within different cores. Of course, they recognize that this ideal global climate δ¹⁸O signal will often not be present. Local weather, changes in the rates at which seafloor sediments are deposited, and disturbances to the sediments after they were deposited on the ocean floor can all distort the original signal. Nevertheless, they believe that it theoretically should be present in all seafloor-sediment cores.

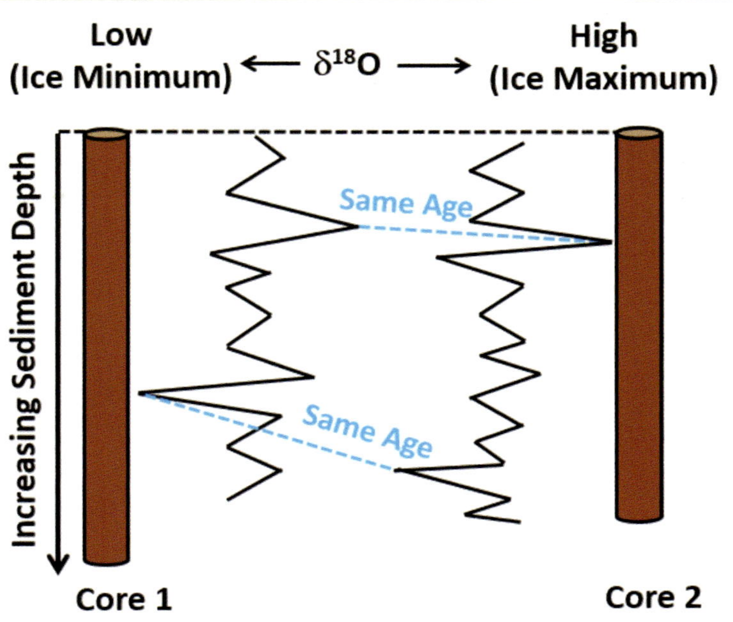

Figure 9.5. Because uniformitarian scientists think that the oxygen isotope signal is a *global* climate indicator, they think that similar $\delta^{18}O$ features in different sediment cores should theoretically be the same age. They use this assumption to transfer ages from core to core using a process called wiggle matching.

This means that if uniformitarian scientists can somehow assign an age to a prominent $\delta^{18}O$ feature (such as a high peak or deep trough) in one core, they think they can theoretically transfer this age to a (presumed) corresponding $\delta^{18}O$ feature within another sediment core (Figure 9.5). This process is called *wiggle matching*. This wiggle matching process played a subtle but important role in the Pacemaker paper.

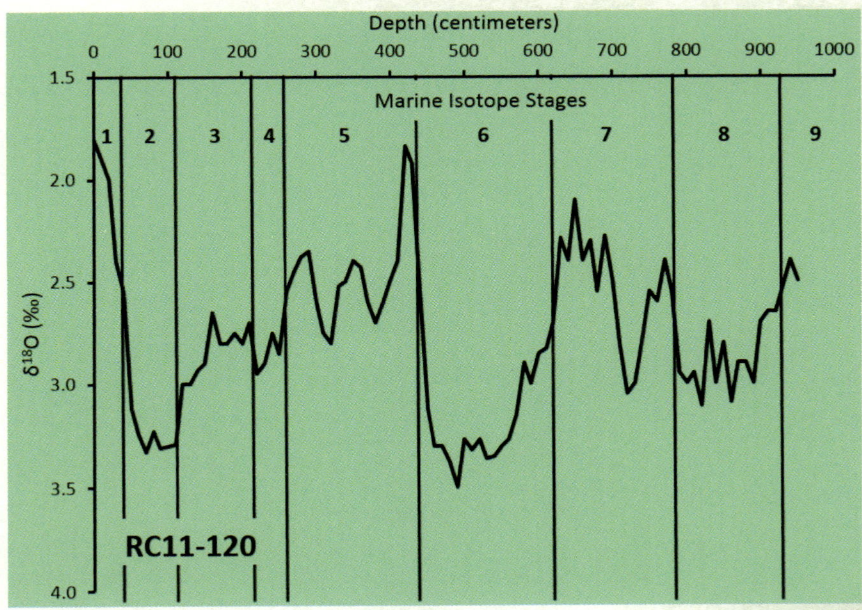

Figure 9.6. The marine isotope stages and their boundaries assigned to the first Indian Ocean core that was used in the Pacemaker paper

Of course, two $\delta^{18}O$ features thought to represent the same climate event in two different cores almost never look exactly alike. This means the person doing the wiggle matching has to make a somewhat subjective determination as to whether these two $\delta^{18}O$ features really represent the same climate event.

Marine Isotope Stages

To help in this wiggle matching process, scientists devised a numbering system involving *marine isotope stages* (MIS). Scientists use marine isotope stages to keep track of similar oxygen isotope features in different sediment cores. Generally (with some exceptions), marine isotope stages with odd numbers represent warm interglacials, while marine isotope stages with even numbers represent cold Ice Ages. Core locations at which $\delta^{18}O$ values are midway between very high or very low $\delta^{18}O$ values are called *marine isotope stage boundaries* (Figure 9.6).

Assigning "Pacemaker" Ages

Before the Pacemaker authors could do their analysis, they had to date the sediments in the two Indian Ocean cores, as well as the $\delta^{18}O$ wiggles within those sediments. Secular scientists do not think they can use radioisotope dating methods to date the deeper seafloor sediments. So, they needed some other way to assign ages to the sediments and $\delta^{18}O$ wiggles.

They used a third deep-sea core, one in the far western Pacific Ocean, for this purpose (Figure 9.2).[17] They thought sediments at that particular location in the western Pacific had accumulated on the ocean floor at a very constant rate for hundreds of thousands of years.[18] Seafloor sediments contain minerals with iron in them. These

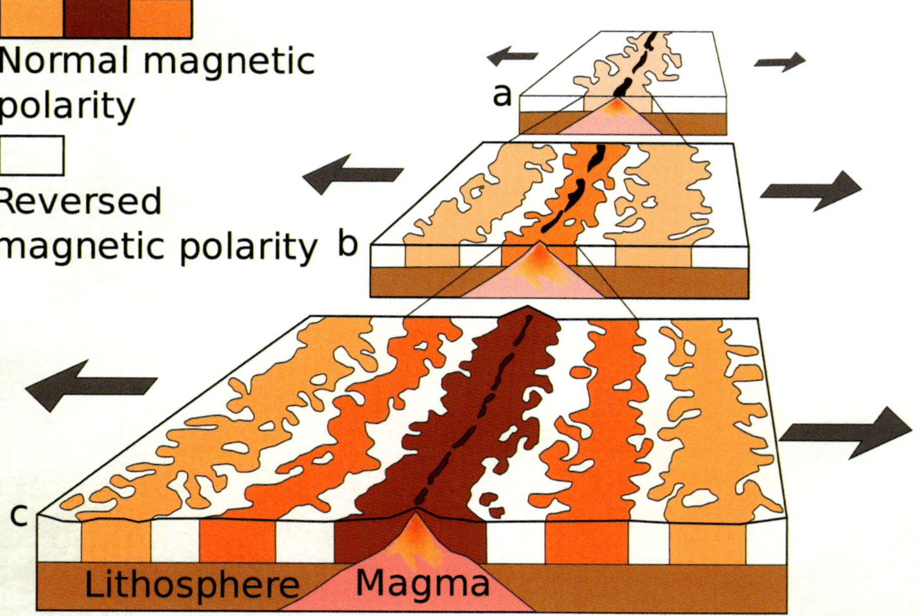

Figure 9.7. Magnetic reversals are depicted in the "striped" pattern seen in volcanic rocks on either side of deep ocean ridges. Volcanic rocks on land can also record magnetic reversals, and the most recent reversal played an important role in convincing secular scientists that the Milankovitch Ice Age theory is correct.

minerals can record information about the direction of the earth's magnetic field at the time those sediments were accumulating. Within this third core they found evidence of a reversal, or flip, of the earth's magnetic field in which the earth's north and south magnetic poles switched places (Figure 9.7).

This western Pacific core, designated V28-238, showed a magnetic reversal at a depth of 12 meters below the core top. Uniformitarian scientists thought this reversal within the core was the most recent magnetic reversal, which they call the *Brunhes-Matuyama magnetic reversal boundary*. We'll call it the M-B reversal, for short. Volcanic rocks on land recorded a magnetic reversal that secular scientists also believed to represent the M-B magnetic reversal. Secular scientists had already used radioisotope dating to date those volcanic rocks as being 700,000 years old. Since the 12-meter-deep sediments and the volcanic rocks were presumably the same age, scientists transferred this age of 700,000 years to those 12-meter-deep sediments.

Because these scientists thought those western Pacific sediments were deposited at a constant rate, and because they thought the age of the core top was zero, they used simple fractions to assign ages to different depths within this western Pacific core (Figure 9.8). For instance, they assumed that the sediments halfway between the core top (at a depth of zero meters) and the magnetic reversal (at a depth of 12 meters) would have an age that was exactly half of 700,000 years. So, they assigned an age of 350,000 years to the sediments midway between those two points, at a depth of 6 meters. They used this method to assign ages to 21 marine isotope stage boundaries within this western Pacific core.

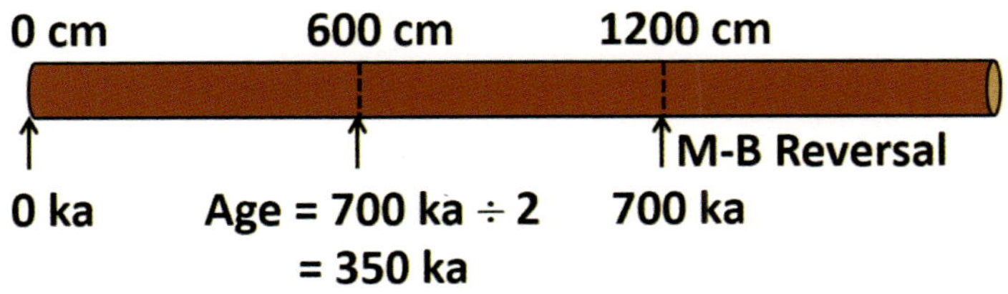

Figure 9.8. Secular scientists used simple fractions to assign ages to different depths within the V28-238 western Pacific core. For instance, the point halfway between the core top and the M-B reversal was assigned an age of 350,000 years—half that of the reversal's assumed age of 700,000 years.

Scientists published these age estimates in a 1973 paper,[19] and the Pacemaker authors used those results in their paper. This is where wiggle matching came into play. Remember that secular scientists think similar-looking $\delta^{18}O$ features, including MIS boundaries,

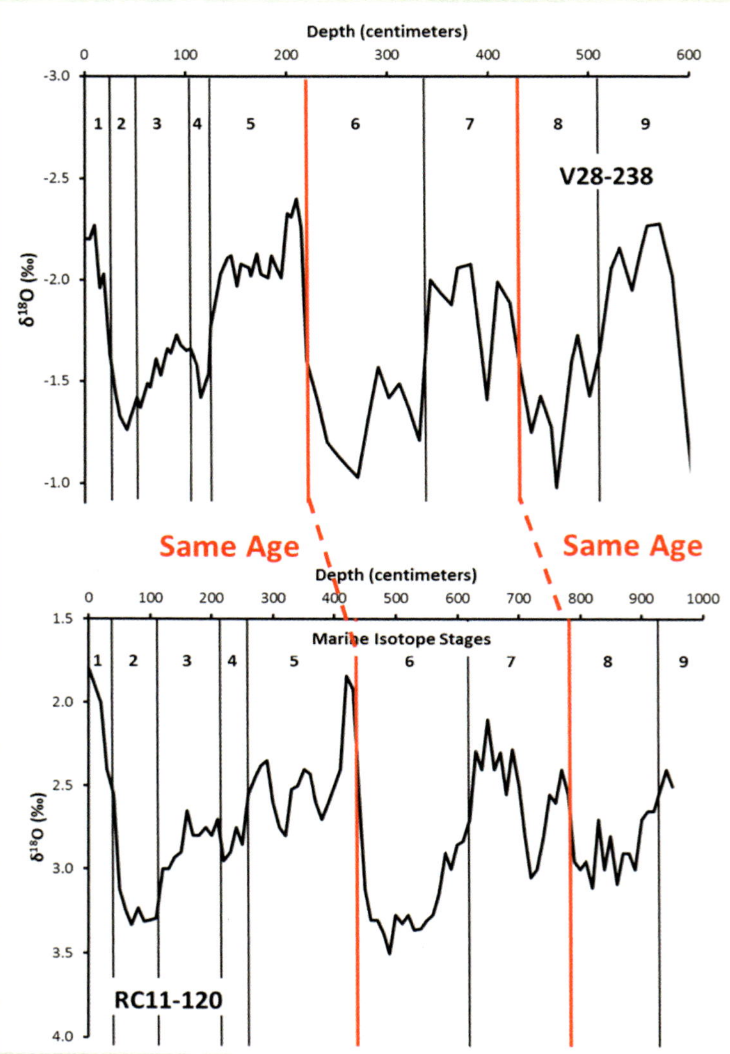

Figure 9.9. Because secular scientists generally think that similar-looking $\delta^{18}O$ features in different cores are the same age, scientists assumed that ages for $\delta^{18}O$ features in the western Pacific core (designated as V28-238) could safely be transferred to similar-looking $\delta^{18}O$ features in the RC11-120 Indian Ocean core. This same process was also used to transfer ages to $\delta^{18}O$ wiggles in the second Pacific core, designated as E49-18.

within different cores should be the same age. So, they transferred three of the 1973 MIS boundary ages to similar-looking $\delta^{18}O$ features in the two Indian Ocean cores (Figure 9.9). They used these three ages as "age control points" to construct age models for the two sediment cores. After doing so, they were able to use a method called *spectral analysis* to examine the data. The results seemed to show support for the Milankovitch theory.

A Big Problem

Creation scientists think these age estimates are vastly inflated. However, there is a huge problem here, even for those who believe in millions of years. The Pacemaker results depended upon an assumed age of 700,000 years for the M-B reversal, but secular scientists now think that the age of this reversal is 780,000 years.[20] This means that secular scientists obtained these iconic results using an age assignment that they themselves no longer accept as valid!

That the accepted age of the M-B reversal is now 780,000 years is common knowledge among scientists who study such things. Even Wikipedia (no friend of the creationist movement) acknowledges the age of this reversal to be about 780,000 years.[21]

This change was big enough to negatively affect the results. If you redo the calculations after taking this revised age into account, then evidence for the Milankovitch theory is greatly weakened, if not completely invalidated. Without getting overly

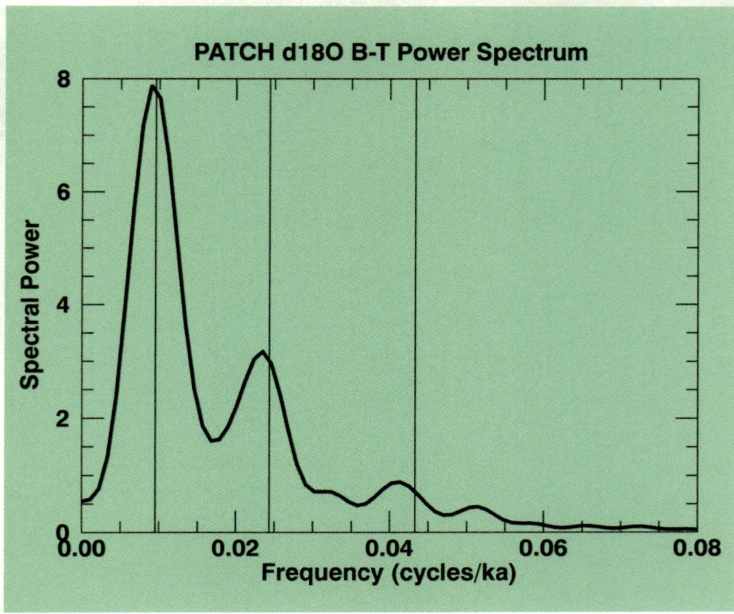

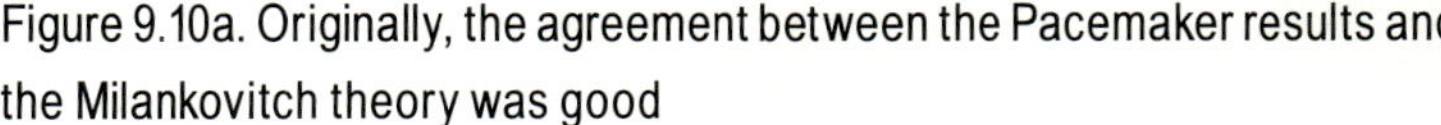
Figure 9.10a. Originally, the agreement between the Pacemaker results and the Milankovitch theory was good

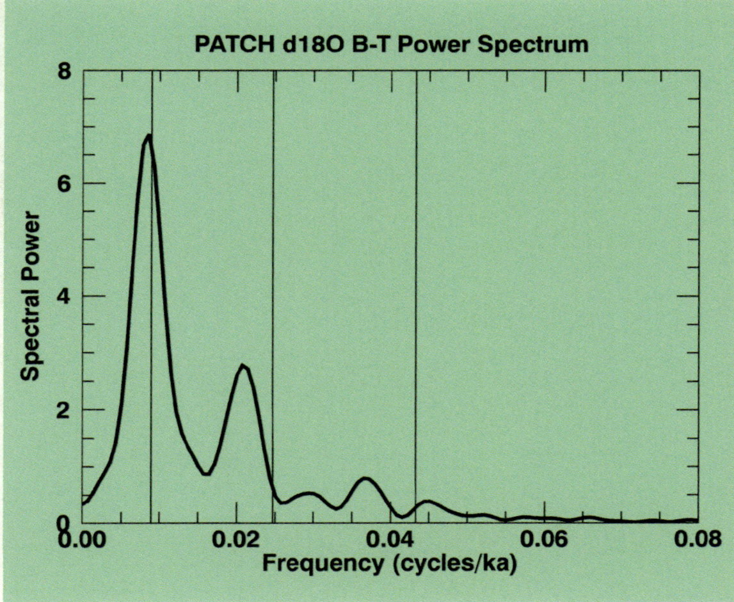

Figure 9.10b. However, the revised age of 780,000 years for the Brunhes-Matuyama magnetic reversal changes the results so they are no longer consistent with Milankovitch expectations

technical, Figure 9.10 illustrates the effect of this change. Graphs like those shown in this figure are called *power spectra*. The peaks on the graph occur at the frequencies of the climate cycles that are making large contributions to the overall climate signal. The vertical lines indicate the eccentricity, obliquity, and precessional frequencies calculated from astronomical data. In Figure 9.10a, the peaks on the graph line up well with the vertical lines, indicating good agreement with the Milankovitch theory. However, in Figure 9.10b, two of the three peaks, those corresponding to periods of 41,000 and 23,000 years, no longer align with the vertical lines, showing that the lengths of the calculated climate cycles no longer agree with the lengths of the astronomical cycles. Furthermore, secular scientists made other changes to the data, and these changes further weaken the results (Figure 9.11).[22]

> “Even Wikipedia (no friend of the creationist movement) acknowledges the age of this reversal to be about 780,000 years.”

Although the first of the three peaks (the one corresponding to a presumed 100,000-year eccentricity cycle) still lines up with Milankovitch expectations, emeritus MIT geophysicist and oceanographer Carl Wunsch has argued that this apparent

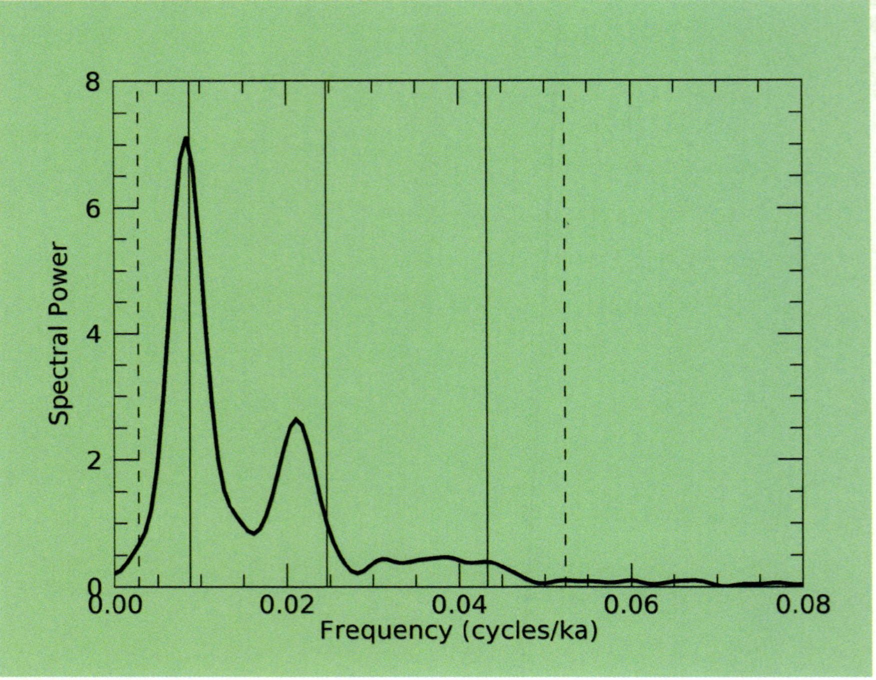

Figure 9.11. Agreement with the Milankovitch theory is even worse when other changes made to the data are taken into account. Dashed lines indicate the frequencies of other astronomical cycles that showed up in the analysis but which were probably too weak to noticeably influence climate

100,000-year cycle could be *stochastic*, meaning it is caused by the random nature of the data, and *not* a result of a 100,000-year orbital cycle.[23,24] So, if two out of three of the climate cycles aren't lining up with Milankovitch expectations, and the one cycle that *is* lining up could be a fluke, then is there any evidence at all for the theory?

Why Has This Gone Unnoticed?

Why don't more scientists seem to be aware of this problem? There is a simple reason. The Pacemaker authors used the 1973 results to assign ages to $\delta^{18}O$ wiggles in the two Indian Ocean cores. As described above, those 1973 results assumed that the age of the M-B reversal boundary was 700,000 years. The Pacemaker authors used these 1973 results and cited the 1973 paper. The Pacemaker authors even stated that these age estimates were calculated using the age of the M-B reversal. But here is the catch: The Pacemaker authors never explicitly stated what that age actually was. Most people who read the Pacemaker paper simply assume that the Pacemaker authors used an age of 780,000 years for the M-B reversal since that is the currently accepted age. But a careful reading of both the 1976 Pacemaker paper and the 1973 paper clearly shows that this was not the case.

For someone with a scientific or mathematical background, these results were easily obtained. Every physicist and mathematician, and probably most engineers, receives some training in spectral analysis, even as undergraduate students. Although that training usually doesn't include the particular method the Pacemaker authors used, it is not difficult to read up on the method and to write a computer code to perform the calculations. In fact, some commercial spectral analysis software programs have this particular method already built into them. These results are not difficult to replicate, and I encourage others to do so. The numbers in my reconstructed sets of the Pacemaker data are freely available online for anyone who wishes to use them.[10] Oddly enough, the original, unaltered data used in the Pacemaker paper do not seem to be publicly available, despite the great importance of the Pacemaker paper to secular thinking. As far as I know, the numbers in my reconstructed data sets are the closest approximations to the original data that are publicly available.

"But here is the catch: The Pacemaker authors never explicitly stated what that age actually was."

Additional Evidence for the Theory?

Some will object that even if all of this is true, then surely there must be strong additional evidence for the Milankovitch theory since many published scientific papers discuss it. In fact, an internet search for "Milankovitch theory" yields more than 100,000 hits! Given the sheer number of publications on this subject, it is easy to assume that additional evidence for the theory must be strong. However, most of these papers assume that the Milankovitch theory is true, and they are using that assumption to draw some kind of conclusion. There are good reasons to think additional evidence for the theory is weak or even nonexistent.[25,26] I discuss these reasons in Appendix D. There, I also discuss a 1997 paper that might seem to rescue the Pacemaker results and explain why I don't consider this paper to be convincing.[27,28]

Why 50 Pleistocene Ice Ages?

Before discussing why this is so significant, now would be a good time to explain why secular scientists say there were 50 Pleistocene Ice Ages in the last 2.6 million

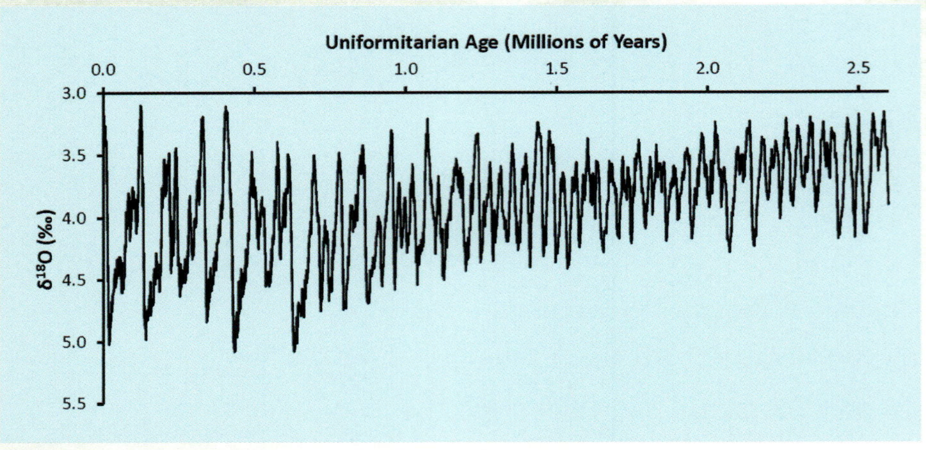

Figure 9.12. Because 50 or so wiggly $\delta^{18}O$ cycles are present in sediments that secular scientists believe to be less than 2.6 million years old, secular scientists claim there were around 50 Pleistocene Ice Ages. Data are from reference 29.

years. As you might guess, this claim is coming from the Milankovitch theory. Once secular scientists convinced themselves that the theory was correct, they started using it to assign ages to other deep-sea sediment cores, as well as to ice cores. Secular scientists call this *orbital tuning*.

Secular scientists used the Milankovitch theory to orbitally tune $\delta^{18}O$ wiggles from dozens of deep-sea cores from around the world. Then they combined these oxygen isotope values into a composite data set they call a *stack*.[29] They think this stacking process reduces uncertainties in the data from the individual cores, giving a more reliable overall signal. If you examine the part of this $\delta^{18}O$ stack said to be less than 2.6 million years old (Figure 9.12), about 50 $\delta^{18}O$ wiggles are visible. (You can count them.) Since secular scientists think the $\delta^{18}O$ values represent global ice volume, they concluded that—*voila*!—there were 50 Pleistocene Ice Ages!

> "Even if one believes that multiple Ice Ages did occur, the most recent glaciers would have bulldozed and destroyed much of the evidence for previous Ice Ages."

That's it! That's the reason for the claim. The claim is *not* based on clear geological evidence such as glacial deposits on land. Even if one believes that multiple Ice Ages did occur, the most

Two polar scientists in red polar clothing working on ice cores over an ice floe

recent glaciers would have bulldozed and destroyed much of the evidence for previous Ice Ages. This would make it impossible to construct a detailed Ice Age history going back millions of years. Rather, the claim of 50 Pleistocene Ice Ages is coming from the Milankovitch theory.

Far-Reaching Implications

The above example demonstrates why the problems with the Pacemaker paper are such a big deal. If evidence for the Milankovitch theory is weak, then this story of 50 Pleistocene Ice Ages is likely wrong, even for uniformitarian scientists who believe in millions of years. What else are secular paleoclimatologists (scientists who study ancient climate) getting wrong?

This demonstrates the way that uniformitarian scientists use the Milankovitch theory to draw conclusions about past climate change. Obviously, if the evidence for the theory is weak or nonexistent, then these conclusions are probably unfounded. And some of these conclusions, as we will discuss in a later chapter, are contributing to climate change alarmism.

I have published a shortcut method you can use to verify the results in Figure 9.10b yourself. Although it takes some effort to understand, it requires only basic high school algebra, finding the equation of a straight line through two points. The Pacemaker paper authors already did the advanced math, so all you need to do is to rescale their results to reflect the change in timescale. If you wish to verify these results on your own, I suggest you get hard copies of both the Pacemaker paper and the 1973 paper the Pacemaker authors referenced, as well as a pencil, paper, and a pocket calculator.[30,31] The method is described in detail in the following online references.[32-34]

References

1. Mann, M. E., R. S. Bradley, and M. K. Hughes. 1999. Northern Hemisphere Temperatures During the Past Millennium: Inferences, Uncertainties, and Limitations. *Geophysical Research Letters.* 26: 759-762.
2. McIntyre, S. and R. McKitrick. 2005. Hockey Sticks, Principal Components, and Spurious Significance. *Geophysical Research Letters.* 32: L03710.
3. Muller, R. Global Warming Bombshell. *MIT Technology Review.* Posted on technologyreview.com October 15, 2004, accessed December 13, 2018.

4. Booker, C. Climate change: this is the worst scientific scandal of our generation. *The Telegraph*. Posted on telegraph.co.uk November 28, 2009, accessed December 13, 2018.
5. Hays, J. D., J. Imbrie, and N. J. Shackleton. 1976. Variations in the Earth's Orbit: Pacemaker of the Ice Ages. *Science.* 194 (4270): 1121-1132.
6. Muller, R. A. and G. J. MacDonald. 2002. *Ice Ages and Astronomical Causes: Data, Spectral Analysis and Mechanisms*. Chichester, UK: Praxis Publishing, xvii. Emphasis added.
7. Ibid, xiv.
8. Hodell, D. A. 2016. The smoking gun of the ice ages. *Science.* 354 (6317): 1235-1236.
9. Maslin, M. 2016. Forty years of linking orbits to ice ages. *Nature.* 540 (7632): 208-210.
10. Hebert, J. 2016. Should the "Pacemaker of the Ice Ages" Paper Be Retracted?—Part 1. *Answers Research Journal.* 9: 25-56.
11. Hebert, J. 2016. Should the "Pacemaker of the Ice Ages" Paper Be Retracted?—Part 2. *Answers Research Journal.* 9: 131-147.
12. Hebert, J. 2016. Should the "Pacemaker of the Ice Ages" Paper Be Retracted?—Part 3. *Answers Research Journal.* 9: 229-255.
13. Hebert, J. 2016. Milankovitch Meltdown: Toppling an Iconic Old-Earth Argument, Part 1. *Acts & Facts.* 45 (11): 10-13.
14. Hebert, J. 2016. Milankovitch Meltdown: Toppling an Iconic Old-Earth Argument, Part 2. *Acts & Facts.* 45 (12): 10-13.
15. Hebert, J. 2017. Milankovitch Meltdown: Toppling an Iconic Old-Earth Argument, Part 3. *Acts & Facts.* 46 (1): 10-13.
16. Wright, J. D. 2001. Cenozoic Climate—Oxygen Isotope Evidence. In *Climates and Oceans*. J. H. Steele, ed. Amsterdam, Netherlands: Academic Press, 316-327.
17. Shackleton, N. J. and N. D. Opdyke. 1973. Oxygen Isotope and Palaeomagnetic Stratigraphy of Equatorial Pacific Core V28-238: Oxygen Isotope Temperatures and Ice Volumes on a 10^5 Year and 10^6 Year Scale. *Quaternary Research.* 3: 39-55.
18. Shackleton, N. J., A. Berger, and W. R. Peltier. 1990. An Alternative Astronomical Calibration of the Lower Pleistocene Timescale Based on ODP Site 677. *Transactions of the Royal Society of Edinburgh: Earth Sciences.* 81 (4): 251-261.
19. Shackleton and Opdyke, Oxygen Isotope and Paleomagnetic Stratigraphy.
20. Shackleton et al, An Alternative Astronomical Calibration.
21. Brunhes-Matuyama reversal. Posted on wikipedia.org, accessed December 13, 2013.
22. Hebert, J. 2018. The "Pacemaker of the Ice Ages" Paper Revisited: Closing a Loophole in the Refutation of a Key Argument for Milankovitch Climate Forcing. *Creation Research Society Quarterly.* 54: 133-148.
23. Wunsch, C. 2003. The Spectral Description of Climate Change Including the 100 ky energy. *Climate Dynamics.* 20: 353-363.
24. Wunsch, C. 2004. Quantitative estimate of the Milankovitch-forced contribution to observed Quaternary climate change. *Quaternary Science Reviews.* 23: 1001-1012.
25. Hebert, J. 2017. A broken climate pacemaker?—part 2. *Journal of Creation.* 31 (1): 104-110.
26. Hebert, Milankovitch Meltdown, Part 3.
27. Raymo, M. E. 1997. The timing of major climate terminations. *Paleoceanography.* 12 (4): 577–585.
28. Hebert, J. 2019. Have uniformitarians rescued the 'Pacemaker of the Ice Ages' paper? *Journal of Creation.* 33 (1): 110-118.
29. Lisiecki, L. E. and M. E. Raymo. 2005. A Pliocene-Pleistocene stack of 57 globally distributed benthic $\delta^{18}O$ records. *Paleoceanography.* 20: PA1003.
30. Hays et al. Variations in the Earth's Orbit. Copies may be purchased at sciencemag.org
31. Shackleton and Opdyke, Oxygen Isotope and Palaeomagnetic Stratigraphy. Copies of this paper may be purchased at cambridge.org
32. Hebert, J. 2017. Testing Old-Earth Climate Claims, Part 1. *Acts & Facts.* 46 (11): 10-13.
33. Hebert, J. 2017. Testing Old-Earth Climate Claims, Part 2. *Acts & Facts.* 46 (12): 10-13.
34. Hebert, J. 2017. A broken climate pacemaker?—part 1. *Journal of Creation.* 31 (1): 88-98.

10 The Genesis Flood Explains the Ice Age

Summary: The creation model provides an explanation for the Ice Age that is vastly superior to the Milankovitch theory. Strong geological evidence exists for only one Ice Age, which happened shortly after the Genesis Flood. The acrostic HEAT can be used to remember key components of this mechanism. An episode of catastrophic plate tectonics occurred during the Flood. Hot magma emerged at the mid-ocean ridges, rapidly forming a brand new seafloor. The heat from this new seafloor resulted in (1) Hot Oceans. The heat from the oceans caused a lot of (2) Evaporation from the ocean surface. The evaporation resulted in much more snowfall at higher latitudes and elevations. The tectonic plate activity caused volcanoes to erupt, putting large numbers of (3) Aerosols (tiny droplets and particles) into the atmosphere. These aerosols reflected sunlight back into space, resulting in cooler summers. Over (4) Time thick ice sheets formed. It is the Bible's short timescale that is critical to allowing this mechanism to work.

Retired National Weather Service meteorologist Michael Oard formulated a creation-based Ice Age model more than 30 years ago.[1] This theory is vastly better than uniformitarian theories. It provides a straightforward explanation for the Ice Age and explains things that are still a mystery to uniformitarian scientists, such as the extinction of the woolly mammoths.

Michael J. Oard

Ice Age Basics

As we discussed in chapter 8, there are two requirements for an Ice Age. First, summers must be colder to keep winter snow and ice from melting. These colder summers must persist for many years. This allows snow and ice from consecutive winters to accumulate, forming large ice sheets. Second, winter snowfall must be heavy, because light snowfall tends to melt, even during colder summers.

Providing Heavy Snowfall

The Genesis Flood provides both of these things. Rapid seafloor spreading occurred at the mid-ocean ridges during the Genesis Flood (chapter 5). The heat from this hot, new seafloor greatly warmed the oceans. If the waters from the "fountains of the great deep" (Genesis 7:11) originated from deep within the earth's interior, these waters would have been quite warm and also may have helped warm the oceans.

The geologic upheaval of the Flood would have thoroughly mixed the waters, causing the oceans to be warm at all depths and latitudes. These warmer oceans caused much greater evaporation from the ocean surface. That in turn placed more moisture into the atmosphere. This resulted in much rain. It also resulted in much snow at higher latitudes and on mountaintops.

The amount of this snowfall would likely have been enormous. Atmospheric scien-

tist Dr. Larry Vardiman and computer scientist Dr. Wes Brewer have used standard meteorology computer programs to estimate the amount of this snowfall. After checking the programs against known meteorological conditions, they simulated much warmer sea surface temperatures. The simulations showed intense snowfall, as expected.[2-5]

Dr. Larry Vardiman

Dr. Wes Brewer

Keeping Snow and Ice from Melting

The necessary summer cooling would be provided by volcanoes. Creation and secular scientists agree that a staggering amount of volcanic activity occurred in Earth's past, activity that dwarfs anything occurring today. Secular scientists think these eruptions occurred over many millions of years. Creation scientists, on the other hand, think these eruptions occurred during the Flood (especially toward the end) and afterward. In fact, creation scientists would argue that today's volcanic activity is a faint echo of this intense volcanism during and after the Flood.

Explosive volcanic eruptions, especially subduction zone volcanoes, can inject large amounts of sulfur-containing gases into the atmosphere. Chemical reactions in the atmosphere produce tiny droplets of sulfuric acid, as well as salts of sulfuric acid. These droplets and particles (known as *aerosols*) reflect much sunlight back into space. They can remain suspended in the stratosphere for two to three years. The 1991 eruption of Mt. Pinatubo in the Philippines, as well as other large eruptions in the recent past, showed that this reduced sunlight results in a slight but noticeable cooling effect. But what if a large number of explosive eruptions occurred at nearly the same time?

That might work, but there is a potential problem. Remember that colder winters are a problem when it comes to explaining an Ice Age. Colder winters cause less snow because the amount of moisture in colder air is much less than in warmer air. However, large explosive volcanoes can provide the needed summer cooling without also making the winters colder. Twentieth-century observations of large explosive eruptions have shown that this is indeed the case. One researcher wrote:

> Several of the largest eruptions are associated with significant drops in summer and fall temperatures, whereas pronounced negative anomalies in winter and spring temperatures are generally unrelated to volcanic activity.[6]

> "Creation and secular scientists agree that a staggering amount of volcanic activity occurred in Earth's past, activity that dwarfs anything occurring today."

There was a small but noticeable drop in global temperatures after the explosive 1991 eruption of Mt. Pinatubo in the Philippines

In other words, these eruptions caused summer cooling but not winter cooling.

It seems volcanoes might really be able to help explain an Ice Age. However, there is another possible problem. Because these aerosols remain in the stratosphere for just a few years, the summer cooling doesn't last very long. The summer cooling for an Ice Age needs multiple large eruptions, occurring over many years.

The Genesis Flood can provide these eruptions. After the Flood, sporadic volcanic activity continued for many years, with gradually decreasing intensity (Figure 10.1). This post-Flood volcanic activity kept injecting aerosols into the stratosphere for many years, allowing these cooler summers to continue.

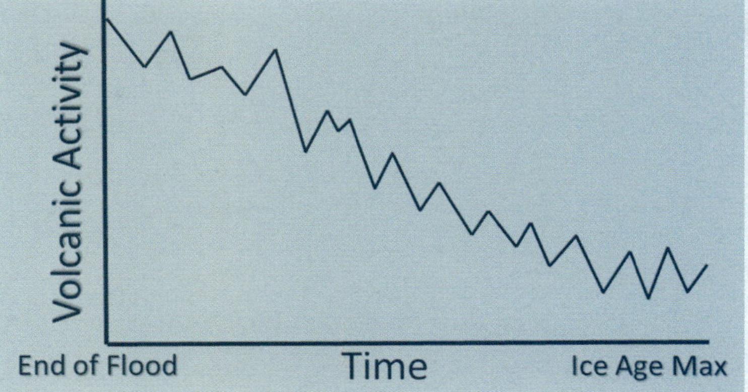

Figure 10.1. Sporadic but gradually decreasing volcanic activity would continue to put aerosols into the stratosphere for many years after the Flood

The Model in a Nutshell

One can remember the key points of this model with the acrostic HEAT. **H**ot oceans would have resulted in increased **E**vaporation from the oceans, which would have resulted in much greater snowfall. **A**erosols from explosive volcanic eruptions provided the cooler summers needed to keep this snow and ice from melting (Figure 10.2). Over **T**ime, thick ice sheets would have grown.

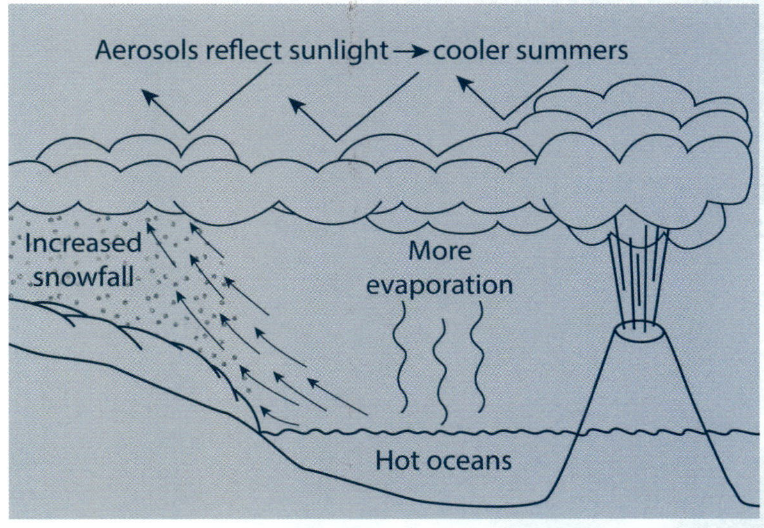

Figure 10.2. An illustration of how hot oceans and volcanic aerosols could cause an Ice Age

How Much Ice?

Some people are under the impression that snow and ice were everywhere during the Ice Age, but that isn't the case. Although some uniformitarian scientists claim the entire earth's surface was frozen in the distant past (the Snowball Earth scenario), the evidence for this claim is weak, as discussed in chapter 7.

Today, snow and ice cover about 10% of Earth's land surface. Both creation and secular scientists agree that during the most recent Ice Age (which creationists say was the *only* Ice Age), snow and ice covered about 30% of the land. Large areas remained ice-free, especially at lower latitudes. Glaciers did not extend down to the Middle East, where nearly all biblical history occurred. In fact, the increased rainfall at lower latitudes during the Ice Age may help explain why portions of the Middle East that are very arid and dry today are described in Scripture as being "well watered" (Genesis 13:10) and as "a land flowing with milk and honey" (Exodus 3:8).[7,8]

Most secular scientists think the North American and European ice sheets were thousands of meters thick. They think these ice sheets had many thousands of years to grow, so they tend to assume that the ice sheets would grow very thick over such long times. Likewise, because the ice sheets in Greenland and Antarctica are very thick, they tend to assume that continental ice sheets in the Northern Hemisphere were also very thick. However, there is strong evidence that the Laurentide, Scandinavian, and British-Irish ice sheets were relatively thin.[9,10] Thinner ice sheets are expected in the Flood/Ice Age model.[11,12] Calculations show that the average ice thickness in the Northern Hemisphere would have been about 700 meters. The average ice thickness in the Southern Hemisphere (nearly all of which was in Antarctica) was about 1,200 meters.

How Long Did It Last?

Eventually, the oceans cooled off and the volcanism decreased sufficiently for the Ice Age to end. Based on heat balance calculations, Michael Oard estimates that the post-Flood Ice Age probably lasted 700 years—500 for the ice sheets to reach their

Figure 10.3. Scientists drill and extract long ice cores from the Greenland and Antarctic ice sheets

> “The Flood/Ice Age model requires intense volcanic activity—and we do, in fact, see evidence of intense volcanism in the Ice Age parts of these ice cores.”

largest sizes and another 200 years for them to melt back to their current sizes.[13] In any case, the Ice Age was a fairly short event. It did not last tens of thousands of years, as claimed by secular scientists.

Evidence for Intense Ice Age Volcanism

The large ice sheets of Greenland and Antarctica started to form during the Ice Age. Secular scientists have drilled and extracted long cores (Figure 10.3) from the ice sheets, some of which go all the way down to bedrock.

The Flood/Ice Age model requires intense volcanic activity—and we do, in fact, see evidence of intense volcanism in the Ice Age parts of these ice cores. Secular scientists think the Ice Age part of one well-known Greenland ice core contains chemical evidence of many hundreds of large volcanic eruptions, eruptions large enough to affect global climate.[14] By uniformitarian reckoning, the Ice Age part of this ice core represents about 100,000 years. However, as will be explained in a later chapter, these ice core age estimates are greatly inflated. The overall cooling effect of even hundreds of large explosive eruptions will be negligible if those eruptions are spread out over 100,000 years. But if the Ice Age portion of this core represents just a few hundred years, as claimed by creationists, then those eruptions were a potent mechanism for providing cooler summers after the Flood.

Time Is the Issue

This illustrates why the age of the earth is not a side issue. The T in the acrostic HEAT stands for *time*, but not just any kind of time. It represents the Bible's short timescale. Many Christians seem embarrassed by the doctrine of recent creation.

However, the Bible's short timescale is the missing key that unlocks the mystery of the Ice Age. Remember, both creation and secular scientists acknowledge that staggering amounts of volcanism have occurred in the past. Likewise, they both acknowledge that large explosive volcanic eruptions can cause summer cooling. Then why can't uniformitarian scientists make better use of volcanoes to explain an Ice Age? It is because they insist that these eruptions were separated by millions of years. Those millions of years would completely dilute any cooling volcanoes might provide. Volcanic cooling helps explain an Ice Age, but only if the eruptions occur close together in time. So, the Bible's short timescale isn't a problem for which embarrassed Christians need to sheepishly apologize—it's the solution!

Moreover, the Lord Jesus Hhimself, during His time on Earth, implicitly affirmed the doctrine of recent creation, as evidenced by several of His statements in the gospels.[15] Furthermore, "Jesus Christ is the same yesterday, today, and forever" (Hebrews 13:8). If He affirmed the doctrine during His earthly ministry, then He must still affirm it. And since He is the Creator, wouldn't He know?

References

1. Oard, M. J. 1990. *An Ice Age Caused by the Genesis Flood*. Santee, CA: Institute for Creation Research.
2. Brewer, W. and L. Vardiman. 2010. Numerical Simulation of Precipitation in Yosemite National Park with a Warm Ocean: A Pineapple Express Case Study. *Answers Research Journal*. 3: 23-36.

3. Vardiman, L. and W. Brewer. 2010. Numerical Simulation of Precipitation in Yosemite National Park with a Warm Ocean: Deep Upper Low and Rex Blocking Pattern Case Studies. *Answers Research Journal.* 3: 119-145.
4. Vardiman, L. and W. Brewer. 2010. Numerical Simulation of Precipitation in Yellowstone National Park with a Warm Ocean: Continuous Zonal Flow, Gulf of Alaska Low, and Plunging Western Low Case Studies. *Answers Research Journal.* 3: 209-266.
5. Vardiman, L. and W. Brewer. 2012. Numerical Simulation of Three Nor'easters with a Warm Atlantic Ocean. *Answers Research Journal.* 5: 39-58.
6. Bradley, R. S. 1988. The Explosive Volcanic Eruption Signal in Northern Hemisphere Continental Temperature Records. *Climatic Change.* 12: 221-243.
7. Vardiman, L. and W. Brewer. 2011. A Well-Watered Land: Numerical Simulations of a Hypercane in the Middle East. *Answers Research Journal.* 4: 55-74.
8. Vardiman, L. 2011. A Well-Watered Land: Effects of the Genesis Flood on Precipitation in the Middle East. *Acts & Facts.* 40 (6): 12-15.
9. Oard, M. J. 2016. Evidence strongly suggests the Laurentide Ice Sheet was thin. *Journal of Creation.* 30 (1): 97-104.
10. Oard, M. J. 2017. Non-glacial landforms indicate thin Scandinavian and British-Irish Ice Sheets. *Journal of Creation.* 31 (2): 119-127.
11. Oard, *An Ice Age Caused by the Genesis Flood*, 93-107.
12. Oard, M. J. 2006. *Frozen in Time*. Green Forest, AR: Master Books, 89-90.
13. Oard, *An Ice Age Caused by the Genesis Flood*, 93-119.
14. Zielinski, G. A. et al. 1996. A 110,000-Yr Record of Explosive Volcanism from the GISP2 (Greenland) Ice Core. *Quaternary Research.* 45: 109-118.
15. See Mark 10:6; Mark 13:19; and Luke 11:50-51a.

Front of Eqi Glacier in west Greenland. Even though explosive volcanic eruptions are known to cause summer cooling, secular scientists cannot effectively use them to explain an Ice Age because of their belief that these eruptions were separated by millions of years.

11 Strengths of the Biblical Ice Age Model

Summary: The HEAT model detailed in the last chapter has many advantages over the Milankovitch theory. It can provide the cooler summers, moderate winters, and higher snowfall needed for an Ice Age. It can also explain things like warmer Ice Age winters in Siberia, wet deserts, driftless areas, and young-looking glacial features such as lack of erosion on glacial deposits. Before secular scientists dismiss the creation model, perhaps they should compare it with their own.

The Flood/Ice Age model does a much better job explaining the Ice Age than uniformitarian models. The details of the model are truly impressive. This chapter provides a summary of some of these details. Multiple monographs, books, and technical articles are available for readers who would like to explore this in more depth.[1-3]

The Big Picture

One of the greatest difficulties with secular (uniformitarian) Ice Age explanations is that they seem to require two contradictory conditions. An Ice Age needs many years of colder summers to keep winter snow and ice from melting. However, one would expect a climate characterized by thousands of years of cold summers to also have winters that are very cold—colder than today's winters. This is a problem because there is much less moisture in very cold air. This would result in less snow, not more. Warmer or mild (not cold) winters are needed to produce abundant snowfall, as expected in an Ice Age. But one generally does not expect cold summers to be accompanied by warmer winters. And one definitely doesn't expect cold summers to be accompanied by warmer winters for the tens of thousands of years claimed by secular scientists!

The Flood/Ice Age model avoids this difficulty because much warmer post-Flood oceans provided greatly increased snowfall over large areas—a "snowblitz."[4] Snowfall was so heavy that ice sheets could still form even with some melting of the previous winter's snow and ice. Hence, summer temperatures in the Flood/Ice Age model still need to be cold, but not as cold as expected by uniformitarian models.

> "One definitely doesn't expect cold summers to be accompanied by warmer winters for the tens of thousands of years claimed by secular scientists!"

A Warmer Ice Age in Siberia?

Surprisingly, Ice Age temperatures in Siberia were warmer than they are today. There is strong evidence that millions of woolly mammoths lived in Siberia during the Ice Age. Although the hairy mammoths were well-suited to a cold environment, it is difficult to see how even they could endure Siberia's bitterly cold winters. January temperatures in Siberia can reach -38°C (-36°F)

Mammoth carcass exhumed in Siberia in the spring of 1902 by the expedition sent by the Russian Academy of Sciences

and even colder.[5] In January 2018, the Russian town of Oymyakon experienced a winter temperature of -88°F. That is colder than the average surface temperature on Mars![6] Woolly mammoths probably couldn't endure such bitterly cold temperatures even if they could find sufficient food and water under those conditions. This is one reason many of us have difficulty getting too worked up over global warming—the people of Siberia might appreciate some warming!

Some have suggested that the mammoths migrated to warmer regions during the winter. However, this is very unlikely for reasons explained in the next chapter. Since mammoth migration is unrealistic, one is forced to conclude that Ice Age Siberian winters were warmer than today. Even uniformitarian scientists have reached this conclusion.[7]

However, warmer Ice Age winters in Siberia contradict uniformitarian expectations. Uniformitarians think that average Ice Age global temperatures were colder than today, perhaps by as much as 10°F. In other areas, temperatures could have been as much as 40°F colder.[8] Based on these numbers, it is not unreasonable to think that Ice Age Siberian winters would likely have been even colder than today! So, uniformitarian expectations, based on uniformitarian assumptions, are contradicting evidence that Siberian Ice Age winters were actually warmer.

Moreover, computer simulations generally predict Ice Age glaciation in areas that we know were not ice-covered, such as the coastal lowlands of Siberia, Alaska, and the Yukon in western Canada.[9-11] For instance, a 2005 simulation predicted ice cover over the Taimyr (or Taymyr) Peninsula in Siberia, even though this area was never covered by ice.[12,13]

Mammoth in the ICR Discovery Center for Science & Earth History, Dallas, Texas

But the Flood/Ice Age model explains both warmer Ice Age Siberian winters and the absence of ice in low-lying coastal areas. The post-Flood oceans would have been quite warm. This would also have been true of the Arctic and northern Pacific and Atlantic Oceans. Immediately after the Flood, no sea ice was present in the Arctic Ocean. In fact, these Arctic waters might have been warm enough that you could swim in them.[14] This absence of sea ice means that warm moist air from these oceans could flow onto the land. This would have helped moderate the climate in Siberia much in the same way that warm air from the Pacific Ocean helps moderate the climate of Seattle, Washington, despite its high latitude. It would have also prevented ice from forming in the coastal lowlands.

Disharmonious Associations

During the Ice Age, cold-climate and warm-climate animals often lived alongside one another, such as hippopotami, reindeer, and elephants in England.[15] These so-called "disharmonious associations" were common.[16] An equable climate, one without extremely hot or extremely cold temperatures, can explain the ability of hot- and cold-climate animals to live alongside one another. Remember that the Flood/Ice Age model predicts cooler summers and mild winters. So, creation scientists expect a more equable Ice Age climate, but this is difficult for uniformitarian scientists to explain.

> "During the Ice Age, cold-climate and warm-climate animals often lived alongside one another, such as hippopotami, reindeer, and elephants in England."

Wet Deserts

Areas that are extremely dry today received abundant rainfall during the Ice Age. The American Southwest had over 100 lakes, some of which were extremely large (Figure 11.1). In fact, Utah's Great Salt Lake is the remnant of a larger lake named Lake Bonneville.

These wet deserts pose a problem for uniformitarian scientists. They generally claim these lakes existed during warmer interglacial periods between Ice Ages.[17] However, by uniformitarian reckoning we are in an interglacial right now, yet the American Southwest today is arid and dry. If "the present is the key to the past," as claimed by uniformitarians, then why did these lakes exist in previous interglacials but not today's? Likewise, what is now the Sahara desert received so much abundant rainfall that it was quite lush, filled with abundant vegetation, wildlife, and human settlements.[18]

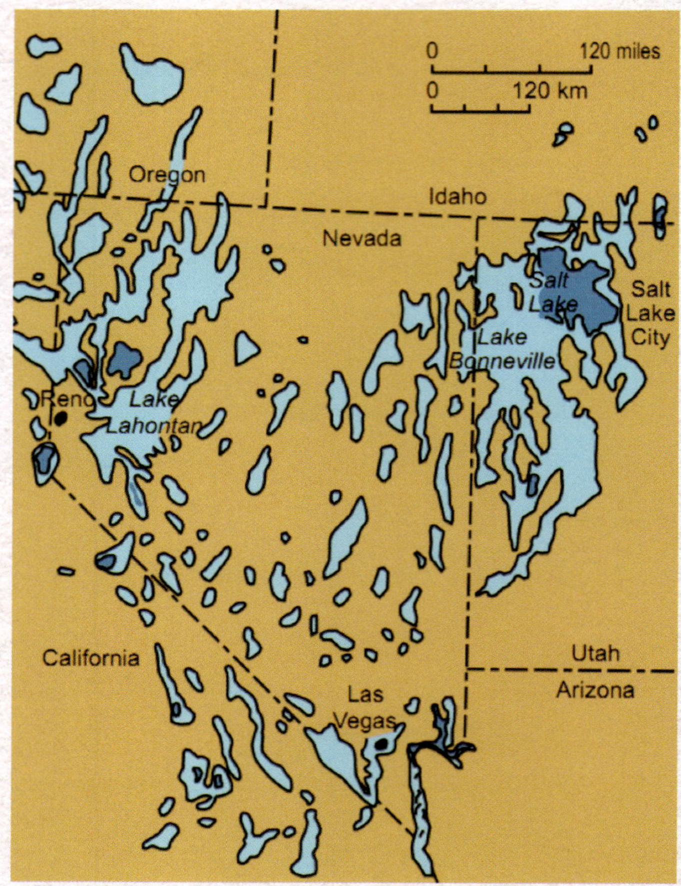

Figure 11.1. Pluvial lakes in the American Southwest are easier to explain in the Flood/Ice Age model. Used by permission of Creation Science Fellowship of Pittsburgh.[19]

It is also worth noting that, according to the Bible, large parts of the Middle East were once more verdant than they are now. Bible scholars think that Job probably lived at roughly the same time as the patriarch Abraham. If so, then he and Abraham may have lived toward the end of the Ice Age. Interestingly, the book of Job has more references to snow, ice, and cold than any other book of the Bible.

> From the chamber of the south comes the whirlwind, and cold from the scattering winds of the north. (Job 37:9)
>
> Have you entered the treasury of snow, or have you seen the treasury of hail? (Job 38:22)
>
> From whose womb comes the ice? And the frost of heaven, who gives it birth? (Job 38:29)

The Bible says that Job lived "in the land of Uz" (Job 1:1), which is associated with the land of Edom elsewhere in Scripture (Lamentations 4:21). Edom was located south of the Dead Sea. Today, the area around the Dead Sea is extremely dry. We know that glaciers did not extend into the Middle East. These references to ice and snow refer to the ice sheets and glaciers at higher latitudes that travelers may have described to Job. Or these verses could be referring to snow and ice that fell in the Middle East during the Ice Age. If the latter, then they are a clue that rain and snowfall were much more

The life of the patriarch Job probably overlapped the time of the Ice Age

The spies sent by Moses returned with grapes from the land of Canaan. More abundant Ice Age rainfall may help explain why the Middle East once received much more rainfall than it does today.

abundant in the Middle East during the time of Job. In fact, by the time of King David, snowfall in the Middle East seems to have become a fairly rare occurrence, so much so that the presence of snow was considered noteworthy (2 Samuel 23:20).

Likewise, the Jordan plain is described as "well watered" in Genesis 13:10, and the land of Canaan is described repeatedly as "a land flowing with milk and honey" (Exodus 3:8, 17; 13:5; 33:3; Leviticus 20:24; Numbers 13:27; Deuteronomy 6:3). Abundant rainfall during the Ice Age may help to explain why the Middle East used to have an abundance of vegetation.

Because the post-Flood Ice Age was marked by abundant precipitation, it is not surprising that areas that today are arid and dry would have received more rainfall in the past. However, explaining why these particular areas received so much rainfall may be difficult because there is still much we don't know about this period.[20]

Driftless Areas

A large area (about 24,000 square miles) in North America was missed by Ice Age glaciers, even though this region was well inside the boundaries of the ice (Figure 11.2). This "driftless area" is located at the corner of southeastern Minnesota, southwestern Wisconsin, northwest Illinois, and northeast Iowa.[21] The terrain within the

Figure 11.2. Is it more likely that 50 thick Pleistocene ice sheets all completely missed the "driftless area," or is more likely that there was only one Ice Age with relatively thin ice sheets?

driftless area is quite rough. Had glaciers covered this area even once, then the terrain would have been greatly smoothed by the advancing glaciers.[22]

For instance, there are sandstone spires within Wisconsin's driftless area that are hundreds of millions of years old by secular reckoning (Figure 11.3). One would expect the glaciers from at least one of the 50 different Ice Ages to have bulldozed and destroyed these spires. This would especially be the case if the ice sheets were very thick, as claimed by some uniformitarian scientists. However, these spires still exist today.

Here is the problem for uniformitarian Ice Age theories. Remember that secular scientists claim that 50 Ice Ages occurred within the last 2.6 million years. Is it reasonable to think that 50 different sets of North American glaciers could somehow all avoid an area of roughly 24,000 square miles, even though ice completely encircled this region (Figure 11.2)?[23] However, if there was really just *one* Ice Age with relatively thin ice sheets as creation scientists argue, then it is much easier to explain why ice never covered the driftless area.

Lack of Erosion for Glacial Features

The features of glacial deposits appear to be relatively young and fresh-looking, as would be expected if the Ice Age occurred just a few thousand years ago. This is not a decisive argument in favor of the biblical model, since the peak of the last Ice Age occurred fairly recently, even by secular reckoning (about 26,500 years ago). However, the relative lack of erosion of these glacial features is consistent with the biblical model.

Missing Ice Age Forests

A strong argument for the post-Flood Ice Age model is the apparent absence of thick Ice Age forests, even in areas *not* covered by ice. Every single living tree in the pre-Flood world was killed during the Flood cataclysm, so the post-Flood Earth would have started out with no forests, and vegetation would have been scanty (Genesis 8:6-12). Since it can take hundreds of years for a naturally occurring forest to grow, one would expect thick forests to have been largely absent during a post-Flood Ice Age lasting hundreds of years.[24]

Scientists use large and small fossilized tree parts such as pollen, seeds, leaves, and wood fragments to infer the locations of past forests.[25] They have been greatly sur-

Figure 11.3. Sandstone spires in the driftless area

> "Since it can take hundreds of years for a naturally occurring forest to grow, one would expect thick forests to have been largely absent during a post-Flood Ice Age lasting hundreds of years."

prised by the apparent absence of thick Ice Age forests, especially in Europe, northern Asia, and North America.[26] For instance, secular researchers have repeatedly noted the dearth of thick Ice Age forests in Europe and northern Asia, despite the fact that their computer models predicted that such forests should have existed.

> Whereas fossil evidence indicates *extensive treeless vegetation* and diverse grazing megafauna in Europe and northern Asia during the last glacial, [computer] experiments combining vegetation models and climate models have to-date simulated widespread persistence of trees.[27]

Other researchers had this to say about Ice Age forests in Europe:

> Reconstructions of the vegetation of Europe during the Last Glacial Maximum (LGM) are an enigma. Pollen-based analyses have suggested that Europe was largely covered by steppe and tundra, and forests persisted only in small refugia. Climate-vegetation model simulations on the other hand have consistently suggested that broad areas of Europe would have been suitable for forest, even in the depths of the last glaciation.[28]

Likewise, thick forests in North America also seem to have been absent during the Ice Age:

> By comparing modern forests and the pollen records they leave behind to pollen records from thousands of years ago, [Margaret] Davis [an ecologist at the University of Minnesota] has created a picture of ancient forests. Her meticulous studies of North America's fossil pollen record show that although trees associated with modern forests existed

> many thousands of years ago, forests as we know them today—dense, continuous stands of trees whose branches form a closed canopy overhead—were likely very rare at the last glacial maximum.[29]

Other researchers have confirmed that Ice Age forests were less extensive than today:

> [North American] tree cover densities during the last glacial maximum were low relative to present, and have increased since.[30]
>
> In the temperate zones of central Europe and the United States where deciduous forests exist today, *vegetation was open and most closely resembled the northern tundra, with grasses, herbs, and few trees during glacial intervals.* Farther south, a broad region of boreal [coniferous] forests with varying proportions of spruce and pine or a combination of both extended almost to the Mediterranean in Europe and northern Louisiana in North America.[31]

The above description seems to contradict the general pattern of "missing" Ice Age forests. However, it should be remembered that using pollen to infer past distributions of forest is difficult. It should also be remembered that different possible rainfall distributions during the Ice Age may have caused forests to grow faster in some areas than in others. For instance, large trees have been found in growth position in Siberian permafrost.[32] In any case, the broad trend is a dearth of thick Ice Age forests, exactly as one would expect after the great Flood.

Uniformitarian scientists sometimes attribute this absence of Ice Age forests to either a cooling trend and/or increased dryness during the Pliocene epoch, before the supposed Ice Ages of the Pleistocene epoch:

> The global cooling that occurred during the Pliocene may have spurred on *the disappearance of forests and the spread of grasslands and savannas.*[33]
>
> Accompanying the general cooling trend of the Pliocene was, as already mentioned, an increased aridity. This led to a number of noteworthy changes in the environment. The Mediterranean Sea dried up completely and remained plains and grasslands for the next several million years. *Another environmental change was the replacement of many forests by grasslands.*[34]

Of course, these hand-waving explanations are not terribly convincing, especially since secular climate and vegetation models suggest that thick Ice Age forests should have been present, especially in unglaciated areas.

> "In any case, the general absence of thick Ice Age forests is extremely puzzling to uniformitarian scientists but is easily explained by the Flood/Ice Age model."

There is evidence that at least some areas of South America, Africa, and Australia were lacking in thick forests during the Ice Age, although this evidence may be too limited to draw firm conclusions.[35-39] Given the difficulty of Antarctic geological field work, data about post-Flood but pre-glaciation trees in Antarctica are very limited. In any case, the general absence of thick Ice Age forests is extremely puzzling to uniformitarian scientists but is easily explained by the Flood/Ice Age model.

References

1. Oard, M. J. 1990. *An Ice Age Caused by the Genesis Flood*. Santee, CA: Institute for Creation Research.
2. Oard, M. J. 2005. *The Frozen Record: Examining the Ice Core History of the Greenland and Antarctic Ice Sheets*. El Cajon, CA: Institute for Creation Research.
3. Oard, M. J. 2006. *Frozen in Time: Woolly Mammoths, the Ice Age, and the Biblical Key to Their Secrets*. Green Forest, AR: Master Books.
4. Ibid, 77-81.
5. Titus, M. Climate in Siberia, Russia. *USA Today*. Posted on traveltips.usatoday.com, updated February 5, 2018, accessed June 7, 2018.
6. Rice, D. So you think you're cold? How does 88 below zero sound? *USA Today*. Posted on usatoday.com January 17, 2018, accessed June 7, 2018.
7. Zukerman, W. Siberia was a wildlife refuge in the last ice age. *New Scientist*. Posted on newscientist.com January 4, 2012, accessed June 8, 2018.
8. Ice Age. *History*. Posted on history.com March 11, 2015, updated June 7, 2019, accessed June 8, 2018.
9. French, H. M. 2017. *The Periglacial Environment*, 4th ed. Hoboken, NJ: John Wiley & Sons, 95.
10. Kaufman, D. S., S. C. Porter, and A. R. Gillespie. 2003. Quaternary alpine glaciation in Alaska, the Pacific Northwest, Sierra Nevada, and Hawaii. In *The Quaternary Period in the United States*, 1st ed. A. R. Gillespie, S. C. Porter, and B. F. Atwater, eds. Amsterdam, Netherlands: Elsevier, 77.

11. Fuller, T. and L. Jackson. 2004. Quaternary Geology of the Yukon Territory. *Yukon Geological Survey*.
12. Zweck, C. and P. Huybrechts. 2005. Modeling of the northern hemisphere ice sheets during the last glacial cycle and glaciological sensitivity. *Journal of Geological Research*. 110: D07103.
13. Zukerman, Siberia was a wildlife refuge in the last ice age.
14. Oard, *Frozen in Time*, 75.
15. Woodward, H. B. and E. T. Newton. 1887. *The Geology of England and Wales: with Notes on the Physical Features of the Country*, 2nd ed. London: George Philip & Son, 514.
16. Lundelius, E. L. Jr. 1989. The implications of disharmonious assemblages for Pleistocene extinctions. *Journal of Archaeological Science*. 16 (4): 404-417.
17. Agenbroad, L. D. et al. Holocene Epoch. Holocene climatic trends and chronology. *Encyclopaedia Britannica*. Posted on britannica.com, accessed June 11, 2018.
18. Tierney, J. E., F. S. R. Pausata, and P. B. deMenocat. 2017. Rainfall regimes of the Green Sahara. *Science Advances*. 3 (1): e1601503.
19. Vardiman, L. 2008. A Proposed Mesoscale Simulation of Precipitation in Yosemite National Park with a Warm Ocean. In *Proceedings of the Sixth International Conference on Creationism*. A. A. Snelling, ed. Pittsburgh, PA: Creation Science Fellowship, 307-319.
20. Oard, *Frozen in Time*, 84-85.
21. About the Driftless Area. National Trout Center. Posted on nationaltroutcenter.org, accessed June 8, 2018.
22. Jefferson, A. The Driftless Area: Fewer glaciers but more topography than the rest of Minnesota. Posted on all-geo.org November 30, 2010, accessed June 11, 2018.
23. News Release: Driftless Area Landscape Conservation Initiative Announces Additional Funding for 2015. Natural Resources Conservation Service, Wisconsin. United States Department of Agriculture. Posted on nrcs.usda.gov March 23, 2015, accessed June 8, 2018.
24. Shubhendu, S. 2015. How to Grow a Forest Really, Really Fast. TED Fellows. Article posted on fellowsblog.ted.com February 18, 2015, accessed October 19, 2018.
25. Birks, H. J. B. and W. Tinner. 2016. Past Forests of Europe. In *European Atlas of Forest Tree Species*. J. San-Miguel-Ayanz et al, eds. Luxembourg: Publication Office of the European Union, 36.
26. Hebert, J. 2019. "Missing" Ice Age Forests: Evidence for the Flood? *Creation Research Society Quarterly*. 56 (1): 48-51.
27. Huntley, B. et al. 2013. Millennial Climatic Fluctuations Are Key to the Structure of Last Glacial Ecosystems. *PLOS ONE*. 8 (4): e61963. Emphasis added.
28. Kaplan, J. O. et al. 2016. Large Scale Anthropogenic Reduction of Forest Cover in Last Glacial Maximum Europe. *PLOS ONE*. 11 (11): e0166726.
29. Not in Kansas Anymore. NASA Earth Observatory. Posted on earthobservatory.nasa.gov August 20, 2002, accessed June 11, 2019.
30. Williams, J. W. 2003. Variations in tree cover in North America since the last glacial maximum. *Global and Planetary Change*. 35: 1-23.
31. Rafferty, J. P. 2011. The Cenozoic Era: Age of Mammals. New York: Britannica Educational Publishing, 194. Emphasis added.
32. Oard, *Frozen in Time*, 141.
33. Pliocene Epoch. Geology Page. Posted on geologypage.com May 5, 2014, accessed October 19, 2018. Emphasis added.
34. Polly, D. The Pliocene Epoch. University of California Museum of Paleontology. Posted on ucmp.berkeley.edu April 30, 1994, updated June 10, 2011, accessed October 19, 2018. Emphasis added.
35. Colinvaux, P. A. et al. 1997. Glacial and Postglacial Pollen Records from the Ecuadorian Andes and Amazon. *Quaternary Research*. 48 (1): 69-78.
36. Tropical rain forest thrived throughout last Ice Age. University of Michigan press release. Posted on news.umich.edu January 5, 2007, accessed October 19, 2018.
37. Mayle, F. E. et al. 2009. Vegetation and Fire at the Last Glacial Maximum in Tropical South America. In *Past Climate Variability in South America and surrounding regions: from the Last Glacial Maximum to the Holocene*. F. Vimeux, F. Sylvestre, and M. Khodri, eds. New York: Springer, 89-112.
38. Monroe, M. H. Ice Age Australia. In Australia: The Land Where Time Began—a biography of the Australian continent. Posted on austrutime.com, updated October 21, 2016, accessed October 19, 2018.
39. Piñeiro, R. et al. 2017. Pleistocene population expansions of shade-tolerant trees indicate fragmentation of the African rainforests during the Ice Ages. *Proceedings of the Royal Society B*. 284: 20171800.

12 Woolly Mammoths in Biblical History

Summary: Another way the creation model is superior to the Milankovitch theory is that it can explain how millions of woolly mammoths thrived in Siberia during the Ice Age. Secular scientists acknowledge that large numbers of woolly mammoths lived in Siberia during the Ice Age, but this contradicts expectations of the secular Ice Age theory since the intense winter cold and lack of food resources should have made this impossible. For woolly mammoths to thrive, post-Flood Siberian winters needed to have been somewhat temperate—something that secular theories cannot explain, yet the creation model does. It can also explain how woolly mammoths went extinct. In more ways than one, the creation model of the Ice Age is superior to secular models.

There is strong evidence that millions of woolly mammoths lived in Siberia, Alaska, and Canada's Yukon during the Ice Age. In fact, millions of woolly mammoths probably lived in just Siberia alone.[1] But the presence of woolly mammoths in Siberia raises questions. First, it is difficult to see how woolly mammoths, even with their thick hair, could have endured the brutally cold Siberian winters. Winter temperatures of -40°F are routine, and temperatures sometimes drop to around -80 to -90°F![2,3] To make matters worse, uniformitarian scientists generally believe the Ice Age was even colder than today![4]

Even if the mammoths could somehow endure these frigid temperatures, it is difficult to see how they could have obtained sufficient food to eat. Adult elephants can eat 200 to 600 pounds of food per day and can drink as much as 50 gallons of water.[5] The same would presumably have been true of woolly mammoths. During the summer, bog vegetation is abundant, but there is very little shrubs and grasses, which mammoths would have likely preferred.[6]

And, of course, obtaining food during the winter would have been even more difficult. During the winter, obtaining sufficient water would also have been problematic since unfrozen water is relatively rare in Siberia.[7] Temperatures would have been more tolerable during the summer, and water would have been more abundant. However, appropriate fodder for the mammoths would still have been a problem.

Some have speculated that perhaps the mammoths migrated out of Siberia during the summer to avoid the brutal winters. However, mammoth expert Gary Haynes noted a number of problems with the migration hypothesis.[8] For one thing, elephant gestation times are typically 18 to 22 months (nearly two years). Assuming that mammoth gestation times were comparable to those of modern-day elephants, such lengthy gestation periods would have made yearly migrations extremely hard on pregnant females. Second, it would have been very difficult for the younger, smaller mammoths, with their much shorter legs, to keep up with the large strides of the adult mammoths. Likewise, the large amounts of food the mammoths needed to eat would have hindered their ability to travel. Worse, the top two or three feet of the permafrost melts during the summer, forming bogs that would make travel hazardous for the mammoths.

"Since annual migration is out of the question, the only logical conclusion for the survival of woolly mammoths is that somehow the Ice Age climate in Siberia was more temperate than it is today."

Since annual migration is out of the question, the only logical conclusion for the survival of woolly mammoths is that somehow the Ice Age climate in Siberia was more temperate than it is today. As noted previously, the Flood/Ice Age model provides a relatively simple explanation for the past presence of woolly mammoths in Siberia. Rapid seafloor spreading and volcanic activity during the Genesis Flood—and possibly warmer waters from the "fountains of the great deep" (Genesis 7:11)—would have significantly warmed the world's oceans. Mixing of these waters during the Flood would have ensured that waters would have been warm at all depths and latitudes. The oceans would have remained warm for hundreds of years.

Moist air from the warm Arctic and North Pacific Oceans would have moderated the Siberian climate. Although winters would still be cold, they would not be nearly as cold as they are today. Winter snowfall in Siberia would have been relatively light. Mammoth expert Dale Guthrie has concluded, based on a number of clues, that during the Ice Age Siberia was a semi-arid grassland, what he called a *mammoth steppe*.[9] Such a grassland would have provided abundant food for the mammoths.

This seems like a fairly simple explanation for how the mammoths could have thrived in Siberia during the Ice Age. So, why can't secular scientists use it? Once again, their belief in millions of years is the problem. Secular scientists think that Arctic sea ice has been present for at least the last 100,000 years.[10] Since secular scientists think most mammoths went extinct about 11,000 years ago, they must somehow explain how the mammoths could have presumably thrived in Siberia for thousands of years when winters in Siberia were as cold as they are today (or even colder). Without the moderating influence of a warm Arctic Ocean, this is very difficult for them to do.

Were Woolly Mammoths on Noah's Ark?

Woolly mammoths likely were not on Noah's Ark, although their ancestors were. Creation researchers think that woolly mammoths, mastodons, and elephants, which belong to the order Proboscidea, are variants of the same basic type, or *kind*, of elephant-like creature (Genesis 1:24-25). So, on Day 6 of the creation week, God did not create a pair of woolly mammoths, a pair of mastodons, a pair of Asian elephants, a pair of Indian elephants, etc. Rather, he created male and female elephantine creatures and put enough potential for variation within their DNA to help them thrive in different potential environments. In fact, some creation researchers think that all the members of the order Proboscidea belong to the same Genesis type or kind.[11,12] This means that the woolly mammoth was simply a particular variant of the elephant kind that adapted to live in a cold environment.

The ability of creatures to adapt to different environments helps answer a fre-

quent skeptical objection to the account of the Genesis Flood and Noah's Ark: How did Noah fit all the different species of animals on the Ark? Skeptics point out that it is absurd to think Noah could have fit all the different species of air-breathing land animals on the Ark. However, creation scientists do not claim that Noah took all the species of air-breathing land animals on the Ark. We claim that he took the air-breathing, land-dwelling *kinds* on the Ark. In theory, a biological species is characterized by the ability to produce fertile offspring, but biologists rarely bother to perform such breeding experiments. Instead, the taxonomic species classification is applied rather arbitrarily, with some biologists using the slightest variation in appearance as justification for naming a new species. This means that, practically speaking, the Genesis kind is not equivalent to a taxonomic species.

Many species actually belong to the same Genesis kind. In fact, creation scientists think that the Genesis kind is roughly equivalent, in most cases, to the family taxonomic level.[13] Hybridization studies have shown that all cats (lions, tigers, cheetahs, house cats, etc.) are the same basic feline type of animal, which creation researchers equate to the Genesis kind.[14] However, even if the Genesis kind were equivalent to the genus rather than the family taxonomic level, Noah would only have needed to take about 16,000 animals on the Ark. Detailed calculations show that the Ark would have provided ample room for these creatures, Noah, his family, and their provisions.[15] If the Genesis kind is more often equivalent to the family taxonomic level, then this would further reduce the number of animals that Noah and his family would have needed to take on the Ark, probably down to a couple thousand or fewer.[16]

The example of the cat family provides us with a good example of how much potential for variation is contained within an organism's DNA, and this quickly refutes the skeptic's ob-

> "In fact, there is much evidence that these changes do not require the deaths of generations of less-fit progeny, as demanded by natural selection, but can instead occur rapidly, even within a single generation!"

jection. Noah did not need to take millions of species on the Ark because the potential for significant morphological variation was already present in the DNA of the representative animals that were on board the Ark.

But this doesn't mean that creation scientists are endorsing biological evolution. Although the members of a created kind carry the potential for much variation, including large variations in size, this variation is always within fixed limits. Reproduction is always after the creature's *kind*. Molecules-to-man evolution requires one basic kind of creature to transform into a completely different kind of creature, but this has never been observed. Creationists would argue that such a transformation is actually prohibited by the information contained within the creature's DNA.

Furthermore, evidence is accumulating that creatures are able to rapidly respond to changing environmental conditions so that their offspring are better equipped to live in new environments. Unlike the random, slow variations expected by evolutionary theory, these changes are rapid, predictable, and targeted to specific environmental conditions. In fact, there is much evidence that these changes do not require the deaths of generations of less-fit progeny, as demanded by natural selection, but can instead occur rapidly, even within a single generation![17-20]

Why Did Mammoths Go Extinct?

As the Ice Age ended, northern high latitude ice melted and flowed into the Arctic

Ocean. Because freshwater is less dense than salt water, this fresh meltwater would have floated on top of the salty seawater. Then it froze, forming a layer of sea ice. With the temperature-moderating effect of these warm oceans now cut off due to the layer of ice, temperatures in Siberia dropped. These colder temperatures would have been too much even for the woolly mammoths to endure, but the slow-moving mammoths would have difficulty moving to a warmer environment. Hence, they went extinct.

However, this temperature drop was not a quick freeze. For many years it has been popular to claim that the woolly mammoths were "quick frozen" as the temperature suddenly dropped to -150°F! The reason for this claim is that some mammoth carcasses contained preserved food content within their stomachs. It was assumed that the only way that this could happen was if the temperature suddenly dropped. We now know that this is not necessary to explain the preserved food in mammoth stomachs, for reasons explained below.

Accompanying this temperature drop was a "drying out" of the climate. Remember that summers during the Ice Age were cooler than today. Yet, the upward movement of warm air allows thunderstorms to develop.[21] Cooler summers would have decreased the number of summer thunderstorms, resulting in a generally drier climate. This effect would not have been significant in the early part of the Ice Age because of the super-abundant moisture in the atmosphere resulting from evaporation from the

Dust storm in Israel

warm post-Flood oceans. However, as those oceans cooled down to their present temperatures, the climate became much drier at the end of the Ice Age.

This temperature drop, combined with drier climates, resulted in the mammoths' extinction, as well as the extinction of many other Ice Age animals. In fact, this drier climate helps explain some important clues about the mammoth carcasses found in Siberia.

Important Clues

Many mammoth carcasses are found in frozen hills of windblown silt called *loess*. This is explained by the Flood/Ice Age model. As the Arctic rapidly cooled off at the end of the Ice Age, the temperature difference between the high northern latitudes and lower latitudes increased greatly. Basic meteorology shows that this greater temperature difference resulted in greater high-altitude pressure difference between the low and high latitudes. This greater pressure difference, combined with the earth's rotation, resulted in strong winds.[22] Because of the dryness of the climate at this time, giant dust storms were the result. Some of the woolly mammoths were overcome by these dust storms and buried alive. After burial, the dust froze around them.

Secular scientists can't use this explanation because it contradicts their uniformitarian climate change story. They believe that the Siberian landscape was already a frozen wasteland with permafrost long *before* the mammoths went extinct. Inserting mammoth carcasses into rock-hard permafrost without destroying them would seem to be impossible. The only logical conclusion is that the mammoths were buried in windblown silt, which froze around them after they had been buried.

As noted earlier, although the extinction of the mammoths involved a drop in temperature, it was not a stupendous temperature drop to -150°F, as some had speculated. Such a huge drop in temperature isn't physically realistic, and it isn't needed to explain the preserved food content found within some mammoth stomachs. Partially preserved food remnants have also been associated with mastodon remains found in climates warmer than that of Siberia.[23-25] Obviously, a quick freeze could not have been responsible for the preservation of those stomach contents. Rather, this preservation is the result of the elephantine digestive system. For elephants and mammoths, most food digestion does not occur in the stomach. Rather, most of the digestion occurs in the cecum and colon after the food has passed through the stomach. Under suitable conditions, these stomach contents can be partially preserved without the need for a stupendous "quick freeze."

References

1. Oard, M. J. 2006. *Frozen in Time*. Green Forest, AR: Master Books, 15-19.
2. Titus, M. Climate in Siberia, Russia. *USA Today*. Posted on traveltips.usatoday.com, updated on February 5, 2018, accessed June 7, 2018.
3. Rice, D. So you think you're cold? How does 88 below zero sound? *USA Today*. Posted on usatoday.com January 17, 2017, accessed May 7, 2019.
4. Ice Age. *History.com*. Posted on history.com March 11, 2015, updated June 7, 2019, accessed June 8, 2018.
5. Elephant Basics. The National Elephant Center. Posted on nationalelephantcenter.org, accessed June 14, 2018.
6. Oard, *Frozen in Time*, 26.
7. Siberia's 'Pole of Cold.' RadioFreeEurope: RadioLiberty. Posted on rferl.org January 19, 2018, accessed June 14, 2018.
8. Haynes, G. 1990. The mountains that fell down: Life and death of heartland mammoths. In *Megafauna and Man—Discovery of America's Heartland*. L. Agenbroad, J. I. Mead, and L. W. Nelson, eds. Hot Springs, SD: Mammoth Site of Hot Springs, South Dakota, Inc., 22-31. Cited in Oard, *Frozen in Time*, 25.
9. Oard, *Frozen in Time*, 29.
10. Johnston, I. Arctic could become ice-free for first time in more than 100,000 years, claims leading scientist. *The Independent*. Posted on independent.co.uk June 4, 2016, accessed May 7, 2019.
11. Oard, *Frozen in Time*, 187-188.
12. Sarfati, J. 2000. Mammoth—riddle of the Ice Age. *Creation*. 22 (2): 10-15.
13. Jeanson, N. 2013. The Origin of Species: Did Darwin Get It Right? In *Creation Basics & Beyond*. Dallas, TX: Institute for Creation Research, 125-131.
14. Pendragon, B. and N. Winkler. 2011. The family of cats—delineation of the feline basic type. *Journal of Creation*. 25 (2): 118-124.
15. Woodmorappe, J. 1996. *Noah's Ark: A Feasibility Study*. Santee, CA: Institute for Creation Research.
16. Jeanson, N. Which Animals Were on the Ark with Noah? Answers in Genesis. Posted on answersingenesis.org May 28, 2016, accessed July 11, 2018.
17. Tomkins, J. P. 2012. Mechanisms of Adaptation in Biology: Molecular Cell Biology. *Acts & Facts*. 41 (4): 6.
18. Guliuzza, R. 2018. Engineered Adaptability: Epigenetics—Engineered Phenotypic 'Flexing.' *Acts & Facts*. 47 (1): 17-19.
19. Guliuzza, R. Fast Evolution Confirms Creationist Theory. *Creation Science Update*. Posted on ICR.org January 16, 2017, accessed July 11, 2018.
20. Thomas, B. Researchers See Fish Adapt in One Generation. *Creation Science Update*. Posted on ICR.org January 17, 2012, accessed July 11, 2018.
21. Lutgens, F. K. and E. J. Tarbuck. 2010. *The Atmosphere: An Introduction to Meteorology*, 11th ed. New York: Prentice Hall, 277-285.
22. Ibid, 168-176.
23. Maugh II, T. H. Remains of Mastodon Reveal Diet: Paleontology: Stomach analysis differs from earlier theories. Bacteria from animal may be 11,000 years old. *Los Angeles Times*. Posted on latimes.com May 4, 1991, accessed May 20, 2019.
24. Lepper, B. T. et al. 1991. Intestinal Contents of a Late Pleistocene Mastodont from Midcontinental North America. *Quaternary Research*. 36: 120-125.
25. Oard, *Frozen in Time*, 155-156.

13 Life During the Ice Age

Summary: The creation model provides clues as to how humans lived during the Ice Age. Life during this time was likely quite pleasant, at least in the Middle East. Some people lived in caves, but they weren't "cave men" in the evolutionary sense. Neanderthals, the iconic cave men, had thick brow ridges and sloped foreheads. Evolutionists call these physical features "primitive," but such features are found in living humans today. Neanderthals during the Ice Age were fully human. The creation model even explains some peculiarities about the book of Job, the events of which likely occurred during the Ice Age. The book of Job has more references to snow and ice than any other book of the Bible. The Ice Age fits well with both biblical and scientific evidence.

As discussed in a previous chapter, life in the Middle East during the Ice Age was likely quite pleasant, due to more abundant rainfall at lower latitudes.[1] But for those living at higher latitudes during the Ice Age, the snow and ice would have made the Ice Age more difficult, but it still would not have been as severe as expected in the uniformitarian model, with its much-colder predicted temperatures.

Were There Cave Men?

The subject of the Ice Age often brings up the subject of "cave men" such as Neanderthals and *Homo erectus*. Although this is not the focus of this book, we discuss it here because of its relation to the Ice Age. Creation scientists argue that these were simply people who lived after the confusion of languages at the Tower of Babel described in Genesis 11:1-9. We argue that evolutionists have erred in portraying these people as primitive ancestors to modern humans and greatly inflated their age assignments.[2]

For example, although the Neanderthals had physical features that today are rare (such as thick brow ridges and sloped foreheads), there is no reason to think they were anything other than fully human. They used tools, wore clothing, hunted large animals, used fire, buried their dead, and in some cases made ornamental objects.[3] Likewise, they had the necessary anatomy required for speech. Even some evolutionists have concluded that they could speak.[4]

Although evolutionists, due to their evolutionary prejudices, long depicted the Neanderthals as sub-human brutes, they are now treating the Neanderthals with much more respect. They now admit that there is simply no evidence that Neanderthals were any less intelligent than we are.[5] In fact, given the deterioration of our genome that has occurred in the last few thousand years, Neander-

Although many artists have depicted the Tower of Babel as a cylindrical structure, most Bible scholars think that it was actually a stepped pyramidal structure called a *ziggurat*

Ziggurat in Ur

Evolutionist reconstruction of a Neanderthal man. Even evolutionists now admit there is no evidence that Neanderthals were in any way inferior to us.

thals were probably smarter than we are, although they did not have our accumulated knowledge and technology.

The Neanderthals sometimes lived in caves, so they technically *were* "cave men." But this in no way implies that they were inferior. The Bible repeatedly mentions people living in caves, such as Lot and his two daughters (Genesis 19:30), David while he was fleeing from Saul (1 Samuel 22:1; Psalm 142:1), the 100 prophets hidden by Obadiah from Queen Jezebel (1 Kings 18:4), and those persecuted by the world (Hebrews 11:38). In fact, Job mentions that some people were living in caves in his day (Job 30:6).

Many Neanderthal skeletons have been found buried in caves, and it is quite interesting to note that this burial practice was apparently common in Abraham's day. Abraham buried his wife Sarah in a cave (Genesis 23:19), and Abraham himself was later buried in that same cave by Isaac and Ishmael (Genesis 25:9).

The End of the Ice Age in Biblical History

As the Ice Age came to an end, the drying out of the climate described in the previous chapter would have occurred. Although it is impossible to say for sure, there may be a connection between this and the multiple famines described in Genesis that occurred (Genesis 12:10; 26:1; 41:56) during the lifetimes of Abraham, Isaac, and Jacob.

At the end of the Ice Age, ice would have rapidly melted, resulting in much catastrophic flooding. For instance, the famous Black Sea flood, incorrectly said by some to have resulted from the Genesis Flood, was the result of catastrophic melting at the end of the Ice Age.[6] The great Lake Missoula Flood in the northwestern United States is another example of catastrophic flooding that occurred at the end of the Ice Age. In fact, even though the evidence for the Missoula Flood is now universally recognized, uniformitarian scientists refused to accept the evidence for this flood for many years because it was so big that it contradicted their uniformitarian expectations.[7] In the following years, even secular geologists have recognized additional evidence for cata-

The burial of Sarah in the cave of Machpelah

During a severe widespread famine, Jacob and his family went to Egypt

strophic Ice Age flooding in North America.[8] So, contrary to popular perception, the biblical record is real history that enables us to make sense of both the geological and archaeological data.

References

1. Tierney, J. E., F. S. R. Pausata, and P. B. deMenocat. 2017. Rainfall regimes of the Green Sahara. *Science Advances.* 3 (1): e1601503.
2. Lubenow, M. L. 2004. *Bones of Contention: A Creationist Assessment of Human Fossils*, rev. ed. Grand Rapids, MI: Baker Books.
3. *Homo neanderthalensis*. Smithsonian Museum of Natural History. Posted on humanorigins.si.edu, accessed November 5, 2018.
4. Hogenboom, M. Neanderthals could speak like modern humans, study suggests. BBC. Posted on bbc.com December 20, 2013, accessed November 5, 2018.
5. Sample, I. Neanderthals were not less intelligent than modern humans, scientists find. *The Guardian*. Posted on theguardian.com April 30, 2014, accessed November 5, 2018.
6. Morris, J. D. 2000. Has Evidence for the Flood Been Found in the Black Sea? *Acts & Facts.* 29 (2).
7. Oard, M. J. 2004. *The Lake Missoula Flood Controversy and the Genesis Flood*. Chino Valley, AZ: Creation Research Society.
8. Ibid, 59-67.

14 Do Deep Ice Cores Prove an Old Earth?

Summary: Some biblical skeptics think that deep ice cores are an unanswerable argument for an old earth. This is because secular age models assign ages of hundreds of thousands of years to deep Antarctic ice cores, and glaciologists claim to have counted more than 100,000 annual layers in deep cores from central Greenland. However, since the age models used for the Antarctic cores implicitly assume millions of years, their vast age assignments actually prove nothing. Likewise, the counting of annual layers in the Greenland cores is much harder than most people realize. Creation and secular age estimates for the top halves of the Greenland ice cores agree to within a factor of three or less, which is reasonable given the difficulty of the counting process. Ages in the bottom halves of the Greenland ice cores are assigned by counting "jumps" in dust content that are thought to represent seasonal cycles. However, a number of factors greatly complicate this process, and creation scientists can plausibly explain secular over-counting. Uniformitarian scientists have themselves demonstrated that the counting process is highly subjective, allowing scientists to manipulate their counts by tens of thousands of years to obtain the "expected" answer.

Scientists have drilled and extracted long ice cores from Greenland. These cores can have total lengths of thousands of meters. One well-known Greenland core is the GISP2 core. Uniformitarian scientists claim that ice near the bottom of this core is more than 100,000 years old.[1,2] They have also drilled and extracted long ice cores from Antarctica. Two of these are the Vostok and EPICA Dome C cores from the Antarctic plateau. Uniformitarian scientists claim that ice near the bottom of the Vostok core is more than 400,000 years old and that ice near the bottom of the EPICA Dome C core is 800,000 years old.[3,4]

Did scientists obtain these ages simply by counting annual layers in the cores? If so, then don't these ice cores prove an old earth? After all, who can argue with simple counting? And if they do prove an old earth, then doesn't this trump all the previous arguments for the Flood/Ice Age model? Although many people are under the impression that the deep ice cores prove an old earth, that is simply not the case.

Vast Time Is Not Needed

The formation of the Greenland and Antarctic ice sheets does not require vast amounts of time. Uniformitarian scientists believe the Greenland and Antarctic ice sheets are millions of years old. However, they believe this because of their general old-earth beliefs, not because millions of years are necessary for the growth of thick ice sheets. Secular scientists acknowledge that thick ice sheets can form in around 10,000 years or less.[5-7] Granted, 10,000 years is more than twice the 4,500 years allowed by the Bible's history. Is this a problem? No, because creation scientists think snowfall was much heavier and more widespread during the post-Flood Ice Age. Much heavier snowfall can plausibly shrink the needed time to less than 4,500 years.

The Greenland ice sheet as seen from space

Figure 14.1. Computer visualization of Antarctica and its surrounding sea ice, constructed from satellite data

Small section of an ice core. Many biblical skeptics mistakenly think deep ice cores are an unanswerable challenge to recent creation.

The *Glacier Girl* World War II airplane was retrieved from beneath 268 feet of Greenland ice that had accumulated in just 50 years

The 1992 recovery of a World War II P-38 Lightning airplane from below the ice of Greenland dramatically illustrates how quickly snow and ice can accumulate under heavy snowfall. In 1942, six P-38 fighters and two B-17 bombers made an emergency landing in southeast Greenland. Fifty years later, aviation buffs successfully recovered and restored one of the P-38 Lightnings. However, the recovery was not easy. Two hundred and sixty-eight feet of snow and ice (mostly ice) had accumulated in just 50 years![8,9] Clearly, snow and ice can accumulate quickly under the right conditions.

However, biblical skeptics had objections to applying this reasoning to the Greenland ice cores.[10] The deep ice cores aren't located in southeast Greenland. Rather, they come from central Greenland, where ice accumulates more slowly. So, the skeptics argued that very rapid snowfall in southeast Greenland, though dramatic, was totally irrelevant to explaining how thick ice could accumulate in just 4,500 years in central Greenland. Admittedly, they had a point. It is invalid to use today's rapid snowfall in southeast Greenland to argue for rapid snowfall in central Greenland, especially when measurements show that central Greenland snow and ice accumulate slowly today.

However, there is a bigger point that the skeptics missed. Remember that the Flood/Ice Age model predicts extremely heavy snow and rainfall even in places that today receive little precipitation. This means that one might expect heavy Ice Age snowfall in central Greenland during the post-Flood Ice Age. This could be the case even though today's snowfall rates in central Greenland are low. The heavy snowfall we see in southeast Greenland today gives us an idea of how quickly snow and ice could have accumulated even in central Greenland during the post-Flood Ice Age. Hence, it is legitimate to use the *Glacier Girl* example to demonstrate how quickly ice and snow can accumulate when precipitation rates are high. Generally, high precipitation rates are expected during the post-Flood Ice Age.

But haven't the vast ages assigned to the ice cores proven that the ice sheets are very old? Even if thick ice sheets can grow in just thousands of years, isn't there strong evidence that the ice sheets are very old? No, but in order to see why, it is necessary to first discuss some ice core basics.

Ice Core Basics

Scientists prefer to drill deep ice cores at geographical features called *ice divides*, where the ice is less likely to have been disturbed. Ice flows outward from the ice divide. In Figure 14.2, ice on the right of the divide flows to the right, while ice on the left of the divide flows to the left. The ice at the divide itself will move neither to the right nor the left. This means that as more ice accumulates, a horizontal layer of ice at the divide can only move straight down. Obviously, the deepest ice cannot continue to move downward because it is blocked by the underlying bedrock.

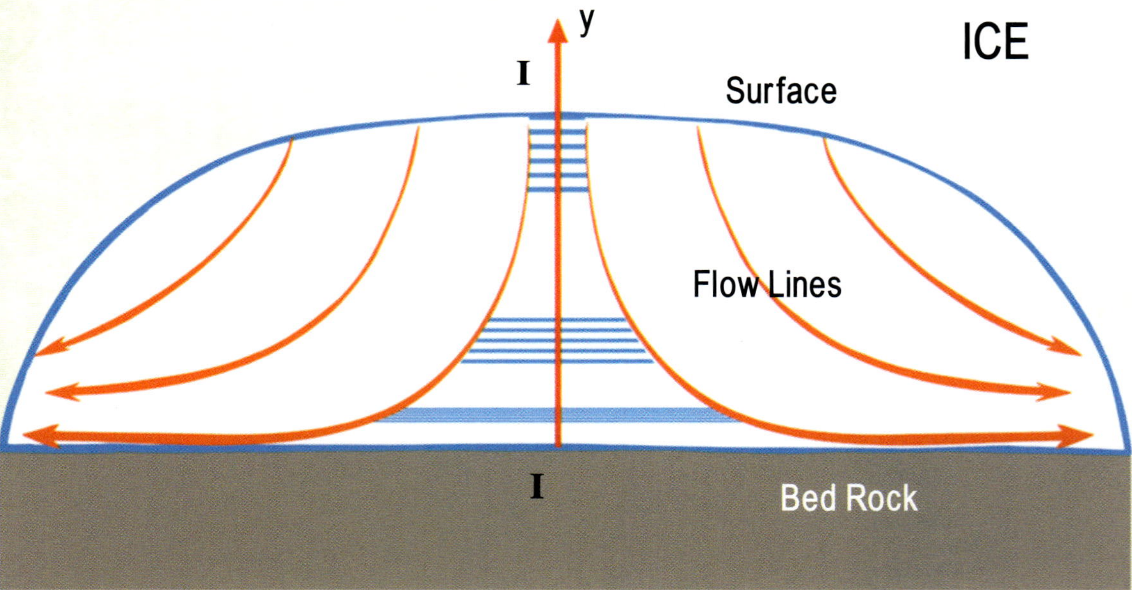

Figure 14.2. Illustration of an idealized ice divide. The ice core would be drilled at the divide at the center of the ice sheet.

For these reasons, if no melting of the ice occurs, then none of these horizontal layers at the divide can disappear. If you could somehow clearly identify annual layers at the ice divide, finding the age of the ice sheet would be trivially easy. Theoretically, every single annual layer since the ice sheet began forming should remain preserved in the ice. These layers will become thinner due to the weight of the overlying ice (Figure 14.3), but the number of these layers should indeed equal the true age of the ice sheet.

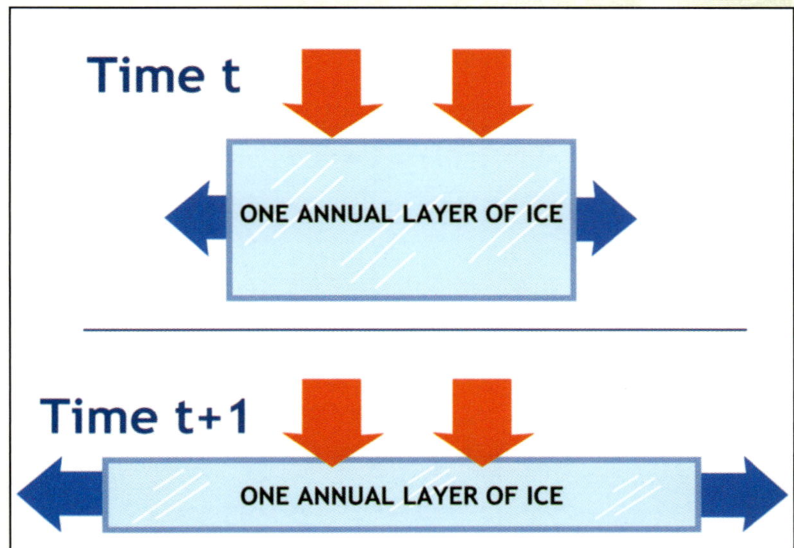

Figure 14.3. Because of the weight of the ice, annual layers become progressively thinner at greater and greater depths within the ice

Complications

In real life, of course, dating the ice isn't that simple. There are complications. Sometimes the ice is disturbed. Sometimes ice at bedrock does melt. A perfect ice divide, like the one described above, doesn't actually exist in real life.

Furthermore, layers aren't always visible. A minimum rate of ice accumulation is necessary for visible layers to consistently appear in a core.[11] However, annual accumulation on the Antarctic plateau, where nearly all the deep Antarctic core sites are located, is very low.[12] This means that scientists can't use simple counting to assign ages to the deep Antarctic cores. Instead, they must use theoretical age-dating models.[13] These theoretical age models are actually the most common method of ice core dating.[14]

Uniformitarian Age Models

Uniformitarian Ice Age models generally assume that the height of the ice sheet has remained constant or nearly constant for most of the time the ice sheet has been in existence.[15] This greatly simplifies the mathematics, but at first glance it seems ridiculous. Whether one believes an ice sheet is just a few thousand years old or millions of years old, everyone agrees that its starting thickness was zero. Creationists and evolutionists can at least agree on that! By treating the thickness of the ice sheet as constant, secular scientist are ignoring the time that the ice sheet was growing in size. Clearly this will be a source of error in their assigned ages. Why do they feel free to do this?

“Neglecting the time for the growth of the ice sheet only makes sense if one believes that the ice is very, very old. It does not make sense if the ice is young.”

Secular scientists know that thick ice sheets can form in just thousands of years. However, they think the ice sheets are millions of years old. They feel free to ignore this error because the thousands of years required for the growth of the ice sheet is a tiny fraction of the millions of years they think the ice sheet has been in existence. After all, if the ice sheets are millions of years old, then what's an error of 10,000 years?

This is why the constant height assumption implicitly assumes millions of years. Neglecting the time for the growth of the ice sheet only makes sense if one believes that the ice is very, very old. It does not make sense if the ice is young.

Thinning of Annual Layers

If one assumes no melting of the ice, the number of annual layers increases by one every year. However, the height of the ice sheet is assumed to remain constant. Also, it is very hard to squeeze a layer of ice into a smaller volume. In this situation, there is only one way that an ever-increasing number of horizontal annual layers can fit into an ice sheet of fixed height. The horizontal layers must become thinner over time (Figure 14.3), with corresponding increases in the surface areas covered by the layers. The amount of thinning is greater at increased depths within the core. Horizontal lay-

Iceberg at sunset, Disko Bay, Greenland

ers at the very bottom will be the thinnest. Secular models predict that thicknesses of annual layers deep in the cores should be quite thin. For instance, a preliminary time-scale for the GISP2 core predicted that the deepest annual layers that could be accurately dated should be four or five millimeters thick.[16,17] The deepest annual layers in the long Antarctic cores can be even thinner, around half a millimeter in thickness.[18] Secular age models predict that annual layers deep in the core become very thin. This will become important a little later in our discussion.

Because visible bands are generally absent from the deep Antarctic ice cores, uniformitarian scientists must use theoretical age models to date the ice. Since these models implicitly assume millions of years, the vast ages assigned to the cores are not surprising. So, the great ages that uniformitarian scientists have assigned to the deep Antarctic cores are actually the easiest for creation scientists to explain. In fact, explaining them is almost trivial.

But What About the Greenland Ice Cores?

On the other hand, the deep cores from central Greenland seem to present a greater challenge to the Bible's short timescale. Scientists did use counting methods to obtain the age of 110,000 years assigned to the bottom of the GISP2 core in central Greenland.[19] Obviously, this age of 110,000 years is much greater than the 4,500 years allowed by the Bible's history. For this reason, some biblical skeptics see the GISP2 core as an unanswerable argument for an old earth.[20]

Secular scientists now prefer to use data from another deep Greenland core called

> "Multiple distinct layers form per year, and the number of layers varies from year to year. Hence, as scientists examine the core, they must make educated guesses as to how many visible bands they should group together and count as one year."

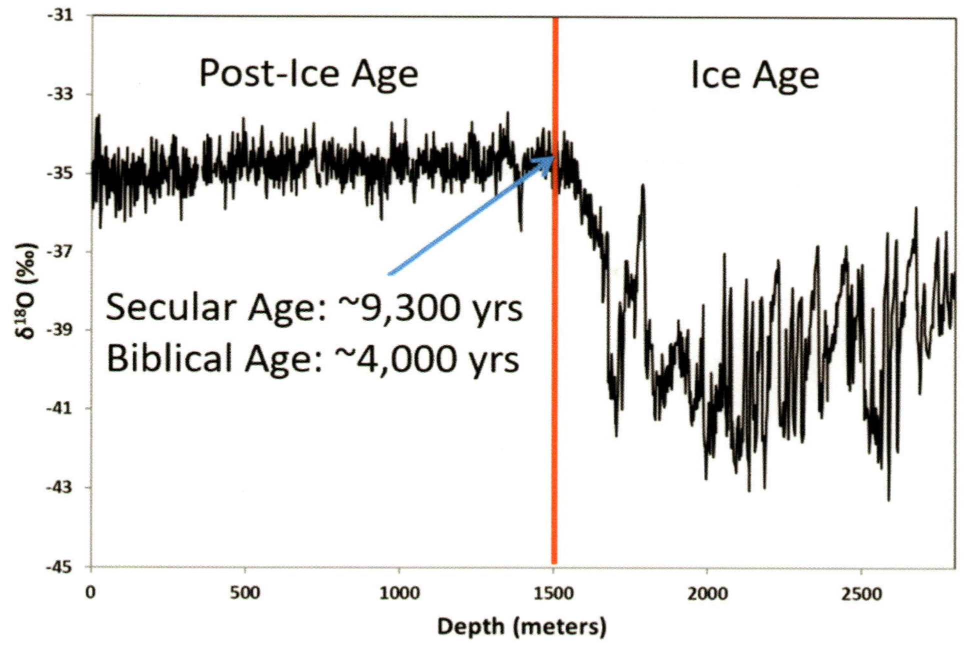

Figure 14.4. Both creation and uniformitarian scientists would agree that a depth of 1,500 meters within the GISP2 ice core marks the transition from the Ice Age to the post-Ice Age climate. Given the uncertainties in the counting process, it is striking that both uniformitarian and creation age estimates for this ice are at least in the same ballpark. Created using data from references 17 and 22.

NGRIP because they think the timescale derived from this newer core is more reliable than the older GISP2 timescale.[21] However, the following discussion of the difficulties in dating the GISP2 core is also applicable to the dating of other deep Greenland cores, including NGRIP.

Dating the Upper Half of the GISP2 Core

Secular scientists assigned an age of 9,300 years to ice at a depth of 1,500 meters in the GISP2 ice core.[22] Based on chemical clues within the ice, both creation and secular scientists think this depth marks the end of the last Ice Age, which creation scientists argue was the only Ice Age (Figure 14.4). A depth of 1,500 meters was about halfway down the GISP2 core. Most people are under the impression that scientists obtained this number because they saw and counted 9,300 visible bands, or perhaps 18,600 bands, one for each summer and one for each winter. That is not the case.

Depth hoar is low-density, large-crystal snow that can form in clear, calm weather under the right weather conditions. Storms can form layers called *wind crusts,* or *wind slabs*, generally consisting of snow that tends to stick together.[23] In fact, scientists have repeatedly observed 15 to 16 storm deposits per year in central Greenland, where the deep ice cores are located.[24] Such conditions can occur repeatedly during the late summer and autumn months. Hence, the number of visible (but not necessarily annual) layers that have formed since the Ice Age probably numbers in the tens of thousands.

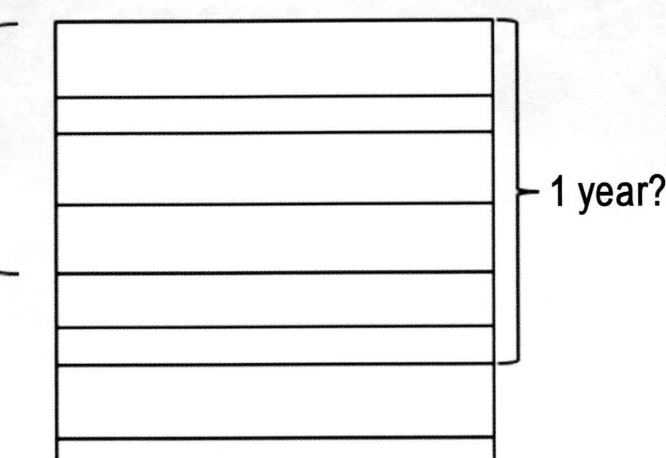

Figure 14.5. Uniformitarian scientists dated the upper half of the GISP2 ice core by counting groupings of layering patterns called *depth hoar complexes*. This required them to make educated guesses about which bands should be grouped together and counted as one year.

A *depth hoar complex* is the collection of all the visible bands deposited within a single year. Here is the important point: Multiple distinct layers form per year, and the number of layers varies from year to year. Hence, as scientists examine the core, they must make educated guesses as to how many visible bands they should group together and count as one year (Figure 14.5). For the upper half of the core, they had about 9,300 such complexes, or groupings, each consisting of multiple numbers of distinct bands.

To put this into perspective, it is worth considering results obtained from the Antarctic coast. Although distinct layers generally are not present in the deep cores from central Antarctica, such distinct layering patterns *are* present in coastal Antarctic ice cores. In order to practice counting layers, scientists will sometimes dig shallow snow pits. One shallow snow pit from the Antarctic coast had around 50 visible layers, but the scientists who studied the layers concluded that they had been deposited over just nine years.[25]

In that light, it is worth comparing this age of 9,300 years with the age that creation scientists assign to it. The Flood happened about 4,500 years ago. In the Flood/Ice Age model, the Ice Age began immediately after the Flood. If we use a round number to make the math a little easier and say that the Ice Age lasted 500 years, this would make the ice at 1,500 meters roughly 4,000 years old.

So, the uniformitarian age of 9,300 years for the end of the Ice Age is in the same ballpark as the age of roughly 4,000 years that creation scientists have deduced for that

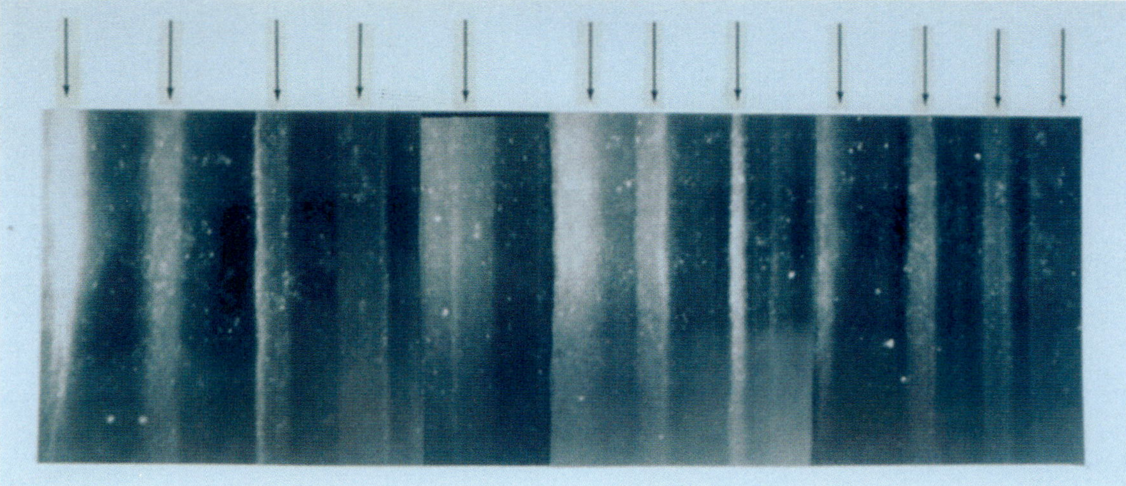

Figure 14.6. Cloudy dust bands from a depth of 1,855 meters in the GISP2 ice core. Uniformitarian scientists assume that these bands represent annual layers even though variations in dust can be caused by factors other than seasonal cycles.

same event (Figure 14.4). Given the difficulties involved in the counting process and the likely very large number of distinct visible bands in the upper half of the GISP2 core, the 5,000 or so additional years that secular scientists have counted is not a problem for the creation position.

In any case, uniformitarian scientists claim that the ice in the bottom half of the core was deposited over 100,000 years, while creationists claim that it was deposited over 500 to 700 years. Secular scientists may be over-counting, but could they really be over-counting by that much? Unfortunately for them, the answer is yes. Creation scientists can plausibly account for the excessive ages assigned to the bottom half of the core.

Dating the Bottom Half of the GISP2 Core

The bottom half of the GISP2 core has much more dust than the top half. Furthermore, there is a seasonal increase in dust concentration during the spring and summer months. Uniformitarian scientists think they can count years or seasons by counting "jumps" in this dust content.[26] Sometimes these increases in dust content are visible and show up as dust-laden cloudy bands (Figure 14.6). However, at greater depths in the core, these cloudy bands are often not visible, so scientists must use instruments to measure the amount of dust in the ice.

Several factors greatly complicate counting peaks of dust content. First, the amount of dust can vary greatly in the bottom half of the core. One respected scientist estimated that the average amount of dust in the deep parts of Greenland cores was about 12 times greater than the amounts in the upper halves.[27] However, dust content can vary by as much as 3 to 100 times greater than the amounts in the upper core sections.[28,29] Why this is important should become apparent shortly.

Second, dust content in ice cores can vary for reasons other than seasonal cycles. Dust concentrations can vary due to storms and can even vary within a single snowstorm.[30,31] This is important, because in an Ice Age lasting 500 to 700 years, one easily expects many thousands of snowstorms.

Third, when cloudy bands are not clearly visible, scientists need to make four to five dust measurements to identify a peak in dust content.[32] But how close together should these four or five measurements be made? The answer to that question depends on how thin you are expecting annual layers of ice to be at a given depth. If you believe that the layers at that depth are very thin, then you will make your measurements very close together. On the other hand, if you believe that the layers are thicker at that particular depth, you can make the measurements farther apart.

But making measurements closer together means that the counter will perceive features in the dust record that he would not perceive at lower resolutions. Peaks and troughs that were not initially visible become apparent at higher resolutions (Figure

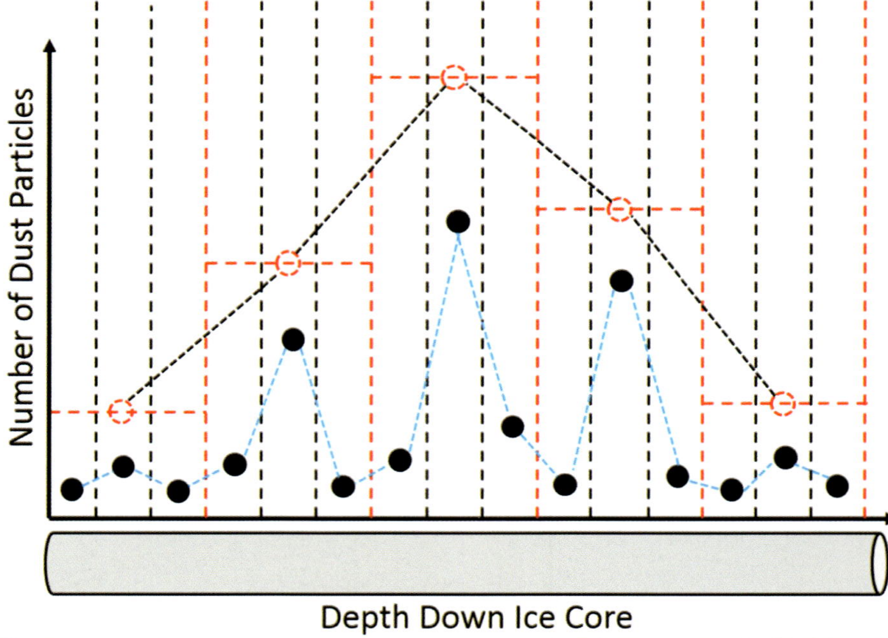

Figure 14.7. The more closely spaced the dust measurements become, the more additional peaks will become apparent

14.7). Should a scientist count these peaks as annual cycles? That depends upon what the counter believes about the past. Remember that uniformitarian models predict that annual layers become very thin at great depths in the ice. These layers will be much thinner than if the ice sheets were young.

So, uniformitarian scientists expect very thin layers at great depths within the ice. But what if the ice sheets have only been in existence for thousands of years? In that case, the true amount of thinning would not be as great as expected by secular scientists. Because uniformitarians expect tens of thousands of annual layers deep within the ice, and because they are making their measurements very close together, it is much more likely that they will mistake short-term jumps in dust content (say, from individual storms) for annual cycles.

GISP2 scientists actually illustrated this very point. They initially assigned an age of 85,000 years to the ice near the bottom of the core, at a depth of 2,800 meters. However, based on Milankovitch expectations and comparison with the Vostok Antarctic ice core, they were expecting the ice at that depth to be 110,000 years old.[33] So two scientists re-counted the dust layers in just the

"Dust concentrations can vary due to storms and can even vary within a single snowstorm. This is important, because in an Ice Age lasting 500 to 700 years, one easily expects many thousands of snowstorms."

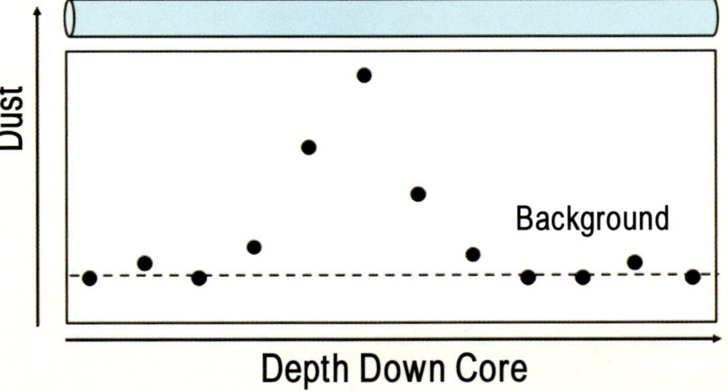

Figure 14.8a. When the background dust level is clear and well-defined, the counting of jumps in dust content is relatively easy

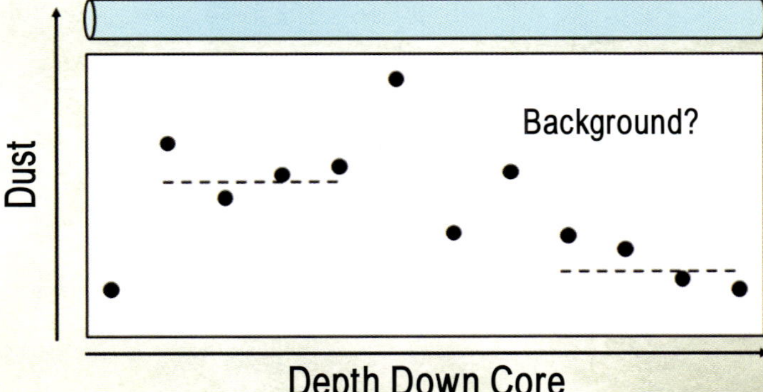

Figure 14.8b. When the background dust level isn't clear and well-defined due to a greatly changing amount of dust, the counting of jumps in dust content is much harder

bottom 500 meters of the core at a much higher resolution (1 mm rather than 8 mm). They then averaged their results, obtaining an age of 111,000 years, which agreed well with the answer they were expecting.[34]

But of course, the only reason they bothered recounting was because they needed an extra 25,000 dust layers in order to get the "right" answer. This illustrates both the subjectivity of the counting process and the relative ease with which scientists can find even tens of thousands of extra "annual" layers when the need arises. The subjectivity of the counting process is further illustrated by the fact that one of the two scientists who did the re-count consistently counted 20% more "annual" layers than the other one![35]

The fact that dust content is highly variable in the bottom halves of the Greenland cores exacerbates the over-counting tendency. If the dust content were more uniform, then it would be easier to determine the "background" amount of dust. This would make identifying seasonal peaks much easier. But because dust concentrations in the core bottoms can vary so dramatically, determining a "true" background level of dust is much harder. This makes it much more likely that scientists may mistake short-term jumps in dust content for seasonal cycles (Figure 14.8).

Other Counting Methods Are Not That Helpful

Scientists used at least five different methods to help them identify visible layers in the ice. Wouldn't using multiple methods help to guard against over-counting? Actually, the multiple methods were not as helpful as one might think. Figure 14.9 shows the counting methods scientists used in different sections of the GISP2 core.[36] The solid arrows indicate a section of the ice where scientists could consistently use a particular method. The dashed arrows indicate sections where scientists could use a method

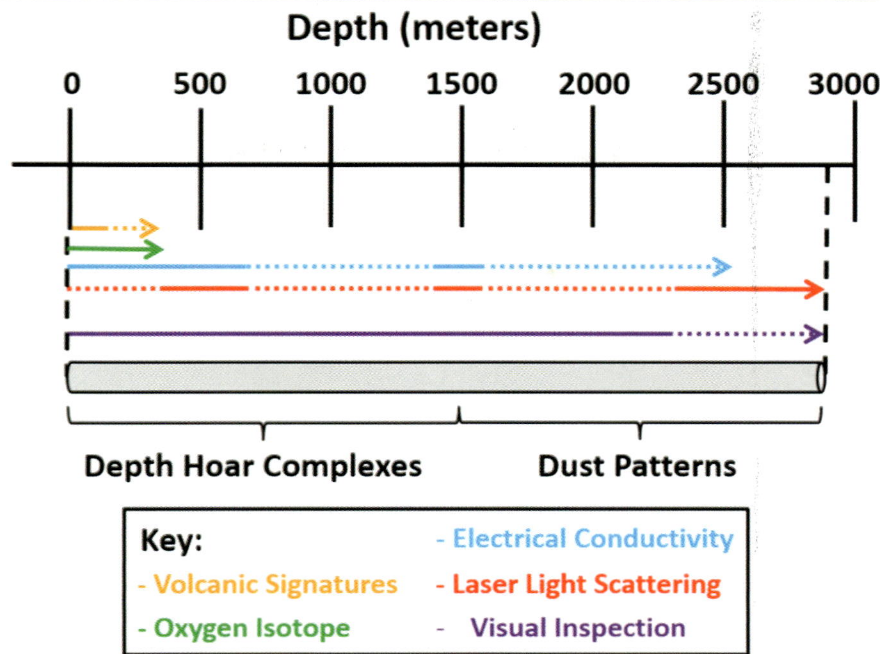

Figure 14.9. The different counting methods used in different sections of the GISP2 Greenland ice core. Solid lines indicate sections where a method could be used consistently, while dashed lines indicate sections where a method could only be used intermittently. Multiple methods could consistently be used as "checks" against one another only infrequently.

only intermittently. For most of the length of the core, scientists were only able to use a single method consistently. This means that these different methods often could not be used as checks on each other.

However, there is an even bigger problem. Even if multiple methods can help you correctly identify patterns in the ice, such as visible bands or increases in dust content, those methods still can't tell you whether the pattern is an annual one or not. So, all the difficulties mentioned above still apply.

Despite popular perception, neither the Greenland nor the Antarctic ice cores provide a compelling argument for an old earth. In this chapter, we critiqued the idea that these ice sheets are millions of years old. In the next chapter, we will present evidence that they are young.

References

1. Meese, D. A. et al. 1997. The Greenland Ice Sheet Project 2 Depth-age Scale: Methods and Results. *Journal of Geophysical Research*. 102 (C12): 26, 411-26, 423.
2. Yau, A. M. et al. 2016. Reconstructing the last interglacial at Summit, Greenland: Insights from GISP2. *Proceedings of the National Academy of Sciences*. 113 (35): 9710-9715.
3. Petit, J. R. et al. 1999. Climate and atmospheric history of the past 420,000 years from the Vostok ice core, Antarctica. *Nature*. 399: 429-436.
4. Lambert, F. et al. 2008. Dust-climate couplings over the past 800,000 years from the EPICA Dome C ice core. *Nature*. 452: 616-619.

5. Weart, S. R. 2008. The Discovery of Global Warming, 2nd ed. *American Institute of Physics*. Posted on aip.org, accessed February 6, 2015.
6. Wilson, R. C. L. et al. 2005. *The Great Ice Age: Climate Change and Life*. London and New York: Routledge and the Open University, 69.
7. Caltech Division of Geological and Planetary Sciences, Class Lecture. Posted on web.gps.caltech.edu, accessed June 1, 2018.
8. Glacier Girl. Website of the P-38 National Association and Museum. Posted on p38assn.org, accessed June 1, 2018.
9. Glacier Girl: The Back Story. *Air & Space Magazine*. Posted on airspacemag.com July 2007, accessed June 1, 2018.
10. Neyman, G. Creation Science Exposed—Greenland Aircraft Claims. Posted on ed5015.tripod.com July 2006, accessed November 5, 2018.
11. Dating a core. Australian Antarctic Division, Department of the Environment and Energy. Revised on antarctica.gov.au June 25, 2002, accessed January 19, 2017.
12. Huybrechts, P. et al. 2000. Balance velocities and measured properties of the Antarctic ice sheet from a new compilation of gridded data for modeling. *Annals of Glaciology.* 30 (1): 52-60.
13. Dating by forward and inverse modelling. Centre for Ice and Climate: Niels Bohr Institute. Posted on iceandclimate.nbi.ku.dk, accessed February 5, 2015.
14. Paterson, W. S. B. 1991. Why ice-age ice is sometimes "soft." *Cold Regions Science and Technology.* 20: 75-98.
15. Cuffey, K. M. and W. S. B. Paterson. 2010. *The Physics of Glaciers*, 4th ed. Burlington, MA: Butterworth-Heinemann, 617.
16. You can easily use a computer spreadsheet and the published age-depth models to estimate average thicknesses of annual layers at different depths.
17. Meese, D. A. et al. 1994. The Accumulation Record from the GISP2 Core as an Indicator of Climate Change Throughout the Holocene. *Science.* 266: 1680-1682. Posted on ncdc.noaa.gov, accessed December 17, 2018.
18. Parrenin, F. et al. 2007. The EDC3 chronology for the EPICA Dome C core. *NOAA*. 3: 485-497. Posted on ncdc.noaa.gov, accessed December 17, 2018.
19. Meese, The Greenland Ice Sheet Project 2 Depth-age Scale. 411-26, 423.
20. Seely, P. H. 2003. The GISP2 Ice Core: Ultimate Proof that Noah's Flood Was Not Global. *Perspectives on Science and Christian Faith.* 55 (4): 252-260.
21. Obrochta, S. P. et al. 2014. Conversion of GISP2-based sediment core age models to the GICC05 extended chronology. *Quaternary Geochronology.* 20: 1-7.
22. The GISP2 timescale (as of May 7, 2019) could be accessed at ftp://ftp.ncdc.noaa.gov/pub/data/paleo/icecore/greenland/summit/gisp2/depthage/gisp2age.txt
23. Wind Slab. Colorado Avalanche Information Center. Posted on avalanche.state.co.us, accessed December 17, 2018.
24. Alley, R. B. 1988. Ice-Core Analysis at Site A, Greenland: Preliminary Results. *Annals of Glaciology.* 10: 1-3.
25. Alley, R. B. 1988. Concerning the Deposition and Diagenesis of Strata in Polar Firn. *Journal of Glaciology.* 34 (118): 283-290.
26. Ram, M. and M. Illing. 1994. Polar ice stratigraphy from laser-light scattering: scattering from meltwater. *Journal of Glaciology.* 40 (136): 504-508.
27. Paterson, Why ice-age ice is sometimes "soft," 75-98.
28. Ibid.
29. Ruth, U. et al. 2003. Continuous record of microparticle concentration and size distribution in the central Greenland NGRIP ice core during the last glacial period. *Journal of Geophysical Research.* 108 (D3): 4098.
30. Jacka, T. H. Antarctic Ice Cores and Environmental Change. *Antarctic Cooperative Research Centre and Australian Antarctic Division*. Posted on chem.hope.edu, accessed January 23, 2017.
31. Personal communication with Michael Oard.
32. Meese et al, The Greenland Ice Sheet Project 2 Depth-age Scale, 411-26, 423.
33. Ibid, 26417-26419.
34. Ibid, 26419.
35. Ibid, 26419.
36. Ibid, 26412.

Greenland Icefjord

15 Positive Evidence for Young Ice Sheets

Summary: Not only is there a lack of evidence that the ice sheets are old, there is positive evidence that they are young and were formed during the biblical Ice Age. Thick ice sheets that secular scientists think began forming millions of years ago cover the Gamburtsev Mountains in east Antarctica. Since these mountains are thought to be about a billion years old, the mountains should have been eroded flat, but they are still rugged and sharp. Additionally, the creation model predicts that many volcanic eruptions happened during the Ice Age that spewed huge amounts of ash and aerosols into the atmosphere. In agreement, cores from both Greenland and Antarctica provide evidence of a large number of Ice Age volcanic eruptions. Also, both the widths of and spacing between volcanic debris layers within the deep Antarctic cores strongly suggest that secular age models are assigning far too much time to the bottoms of those cores.

Explosive volcanic eruptions can place large amounts of sulfur dioxide (SO_2) into the atmosphere. Atmospheric chemical reactions between sulfur dioxide and other chemicals produce sulfuric acid (H_2SO_4) droplets. These droplets can be transported up to the stratosphere, where they can remain for several years. Some of these droplets fall out on the Greenland and Antarctic ice sheets. Scientists can use instruments to detect the presence of these acids in the ice.

Remember that intense volcanism is an important part of the biblical Ice Age model. If this model is correct, then we should expect to see indications of intense volcanism in the Ice Age parts of the ice sheets. And we do! Based on chemical measurements in the GISP2 ice core, uniformitarian scientists estimated that 700 large volcanic eruptions, eruptions large enough to influence the climate, occurred during the Ice

Mount St. Helens eruption, May 1980

Age. Of course, these are only the eruptions that left an acid layer in the ice.[1] In the old-earth model, the effect of these eruptions on the climate is minimal since the eruptions are spread out over 100,000 years. However, if this ice really accumulated over just 700 years, as in the biblical model, then these eruptions would be a potent cooling mechanism. So, these chemical clues in the ice are consistent with the intense Ice Age volcanism predicted by the biblical Ice Age model. Cores from Antarctica also provide evidence of a large number of Ice Age volcanic eruptions:

"So, these chemical clues in the ice are consistent with the intense Ice Age volcanism predicted by the biblical Ice Age model."

> A total of 25 ash bands and an estimated 2000 dust bands are preserved in the Byrd ice cores. These debris bands are estimated to have been deposited on the surface of the ice sheet between 43,500 and 7500 years ago and testify to much more extensive fallout of volcanic ash in Antarctica during the latter part of the Wisconsin than was originally suspected.[2]

The authors defined "dust bands" as cloudy bands that contain fine glassy and crystalline fragments, and "ash bands" as bands of coarser ash particles. So, the authors are saying that around 2,000 fine and coarse ash layers are located in the Ice Age (Wisconsin) part of this Antarctic ice core. Again, creationists expect large amounts of volcanic activity during the post-Flood Ice Age.

Thick Volcanic Ash Layers

As noted above, some volcanic eruptions deposit ash and tiny fragments of volcanic glass called *tephra* onto the ice. These particles generally fall out of the atmosphere even more quickly than stratospheric aerosols, usually in weeks to months. Due to the thinning of annual layers at greater depths, these tephra layers should also become thinner over time. However, creationists would predict less thinning of the tephra layers than would uniformitarians. If the ice sheets are young, then uniformitarian age models are assigning too much time to volcanic tephra layers. Tephra layers would appear to have fallen over the course of several years when in reality they fell over just weeks or months. In this light, it is interesting that uniformitarian scientists concluded that a particular tephra layer in Antarctica's Dome Fuji core represented about five years of ash fallout.

> The thickness and duration at deposition, determined by a simple ice-flow model, suggests that a violent volcanic eruption caused ash to fall onto the Antarctic ice sheet for ~5 years and to form a ~100 mm thick tephra layer at 117 ka BP [117,000 years before present].[3]

The actual measured thickness was 24 mm. However, using their uniformitarian age-depth model, secular scientists estimated that the initial thickness of the ash was ~100 mm when it first fell on the snow. Since stratospheric aerosols generally remain in the stratosphere for about 1.5 years, the authors concluded that the volcanic eruption lasted for more than three years.[4] Furthermore, based on chemical analyses on the tephra, they concluded that this ash layer probably came from a single volcano, most likely Antarctica's Mt. Berlin. The scientists estimated that their timescale was accurate to within 20%, which means the time that the ash was falling out on the ice could have been as little as four years or as many as six. In any case, this would imply that a single volcano kept injecting ash into the atmosphere for at least a year, maybe four years. Is this reasonable?

Science popularizer Bill Nye drew attention to the issue of tephra in ice cores in

the documentary *Bill Nye: Science Guy*, which aired in the United States on the Public Broadcasting Service (PBS) in April 2018. Nye, along with climatologist Dr. James White, visited a scientific station in Greenland at which ice cores were being extracted.[5] Because the documentary was meant to combat those whom Nye deemed "anti-science" (especially creationists), the choice to film at this Greenland station cannot have been a coincidence. Since uniformitarians consider the ice cores to be strong arguments for an old earth, Nye obviously thought that highlighting Greenland ice cores would be a good way to argue against the Bible's short 6,000-year chronology. Yet, he inadvertently demonstrated a key point that creationists have been making for a long time. I described the incident in an ICR news article:

> At the 1:04:53 mark in the documentary, Bill Nye and climatologist Dr. James White are examining a recently extracted Greenland ice core section. Bill Nye notices a dark band of tephra within the ice and asks if it represents a volcanic eruption. Although Dr. White does not respond, it is obvious from the following conversation that the answer is "yes." White states that this ice is about 27,000 years old. Nye then notes that the tephra layer, which appears to be unbroken, with no gaps, represents a time span of 15 to 17 years. Although Nye does not explicitly explain his reasoning, his quick mental calculation is based on the assumption that the nearby visible bands in the ice represent *annual* layers. Since he estimated the tephra layer to be about 15 to 17 times thicker than the nearby "annual" layers, the tephra should represent 15 to 17 years. Nye is obviously very surprised that fallout from

> "That means that if one assumes, as Bill Nye did, that each visible band in the nearby ice is an annual layer, then ash was falling onto the Greenland ice for about 12 years."

> a volcanic eruption would last that long, so he asks, "Is that possible?" White does nothing to indicate that Nye's mental calculation is in error, and he responds that it most certainly *is* possible.[6]

Remember that even very small ash particles should fall out in a few years. Coarser, larger ash particles fall out much more quickly. That means that if one assumes, as Bill Nye did, that each visible band in the nearby ice is an annual layer, then ash was falling onto the Greenland ice for about 12 years. Fallout of volcanic ash lasting that long would truly be astounding!

If one closely examines the video footage, it is clear that there are color and texture variations within the ash layer, suggesting that some of the ash may represent immediate fallout of coarser ash particles while some of it represents later fallout of finer particulates over a longer time period. But even if one were to argue that two-thirds of the thickness of this ash layer was caused by immediate fallout of coarse ash from a nearby volcano, that leaves five years or so of residual fallout of finer ash particles, which would still be an unusually long time for volcanic ash to settle.

Of course, scientists who specialize in the dating of ice cores recognize that not every visible band in the ice is an annual layer. Yet the popular perception is that each visible band (or perhaps every two bands) does represent an annual layer. This particular tephra layer reminds us that this is not the case.

Decreasing Tephra Frequency in the Ice

Tephra in ice cores favors the young-earth model. Uniformitarian scientists are assigning too much time to both the Greenland and Antarctic ice cores, but the timescales for the deep Antarctic ice cores are especially inflated. Whereas the deep Greenland ice cores are only said to represent about 100,000 years, the Vostok, EPICA Dome C, and Dome Fuji deep Antarctic ice cores are said to represent 420K, 800K, and 700K years, respectively.

Also, these age models assign nearly all their time to the deepest parts of the ice cores. Remember that uniformitarian scientists assigned about 100,000 years to just the bottom half of the nearly 3,000-meter-long GISP2 ice core in Greenland, but only about 9,300 years to the upper half of the core.

Creation scientists argue that uniformitarian age models are assigning huge amounts of imaginary time to the deepest parts of the Antarctic cores. If this is really the case, then volcanic eruptions in the deepest parts of the Antarctic cores will seem very infrequent. Tephra layers in the bottoms of the cores might represent volcanic eruptions separated by just decades or centuries, but the inflated uniformitarian age scales will make them seem to be separated by hundreds of thousands of years. And this is exactly what is observed.[7] Uniformitarian scientists commented on a striking decrease in frequency of volcanic eruptions that they observed in the Vostok and EPICA Dome C (EDC) Antarctic ice cores:

> A striking feature emerging from our study is that the frequency of visible tephra in the Vostok and EDC cores decreases dramatically in the ice older than ca 220 ka (Fig. 5). The last [most recent] 220-ka sections of both records contain about a dozen discrete tephra layers while only one event is identified at EDC and two at Vostok in the interval 220-414 ka, encompassing more than two complete climate cycles. Tephra layers even disappear from 414 to 800 ka, i.e. the bottom of the EDC core.[8]

The authors observed that this "drastic" decrease in tephra frequency was apparent in the Dome Fuji core, too.[9] The tephra layers in these three cores are plotted as a function of depth in Figure 15.1 and as a function of time in Figure 15.2. When the layers are plotted as a function of depth, they are spaced randomly but more or less evenly throughout the depth of the cores (although they become a little more spread out toward the bottoms). However, when they are plotted as a function of assigned age, the apparent drop in frequency is obvious.

The weight of the ice causes the ice to thin with depth, which will cause the tephra layers to become thin as well. Could it be that the deep ice has thinned so much that the tephra layers simply are no longer visible? No, because two tephra layers *are* visible at the very bottom of the Vostok core (Figures 15.1 and 15.2), where the greatest amount of thinning should have occurred. If very thin tephra layers are visible at the very bottom of the Vostok core, then broader tephra layers higher up in the Vostok

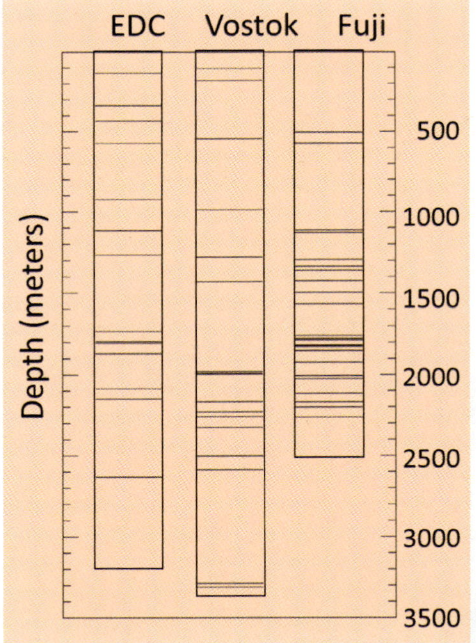

Figure 15.1. The depths at which volcanic tephra (ash) layers are found within the EPICA Dome C, Vostok, and Dome Fuji deep Antarctic ice cores

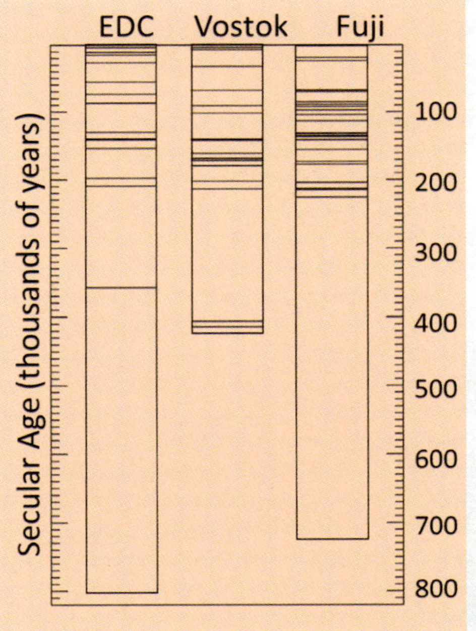

Figure 15.2. The ages that theoretical secular dating models have assigned to these tephra layers in the deep Antarctic ice cores. There is a dramatic apparent decrease in the frequency of these volcanic eruptions in the supposed distant past. This is surprising if "the present is the key to the past," as uniformitarians claim, but it makes perfect sense if these age models are assigning hundreds of thousands of years of fictitious time to the ice core bottoms.

core, if they are present, should also be visible. The fact that these tephra layers are not visible strongly suggests that they just aren't there.

The authors ruled out the ice thinning explanation for this apparent decrease in tephra frequency.[10] Because of the wide geographic separation of the cores, they concluded that this apparent decrease in frequency was real and could not be blamed on factors like changing wind patterns.[11] Of course, this apparent decrease in volcanic frequency is expected if uniformitarian scientists are assigning hundreds of thousands of fictitious years to the bottom sections of the Antarctic ice cores. This apparent decrease in frequency is just a consequence of their exaggerated timescales.

Little Erosion Under the Ice

Two major ice sheets cover Antarctica. Secular scientists say the East Antarctic ice sheet is about 14 million years old, although some claim an ice sheet covering the entire continent formed 34 million years ago.[12,13] They say that the West Antarctic ice sheet is a few million years old, although its size may have varied considerably since then.[14] Since the 1950s, scientists have known that the Gamburtsev Mountains in

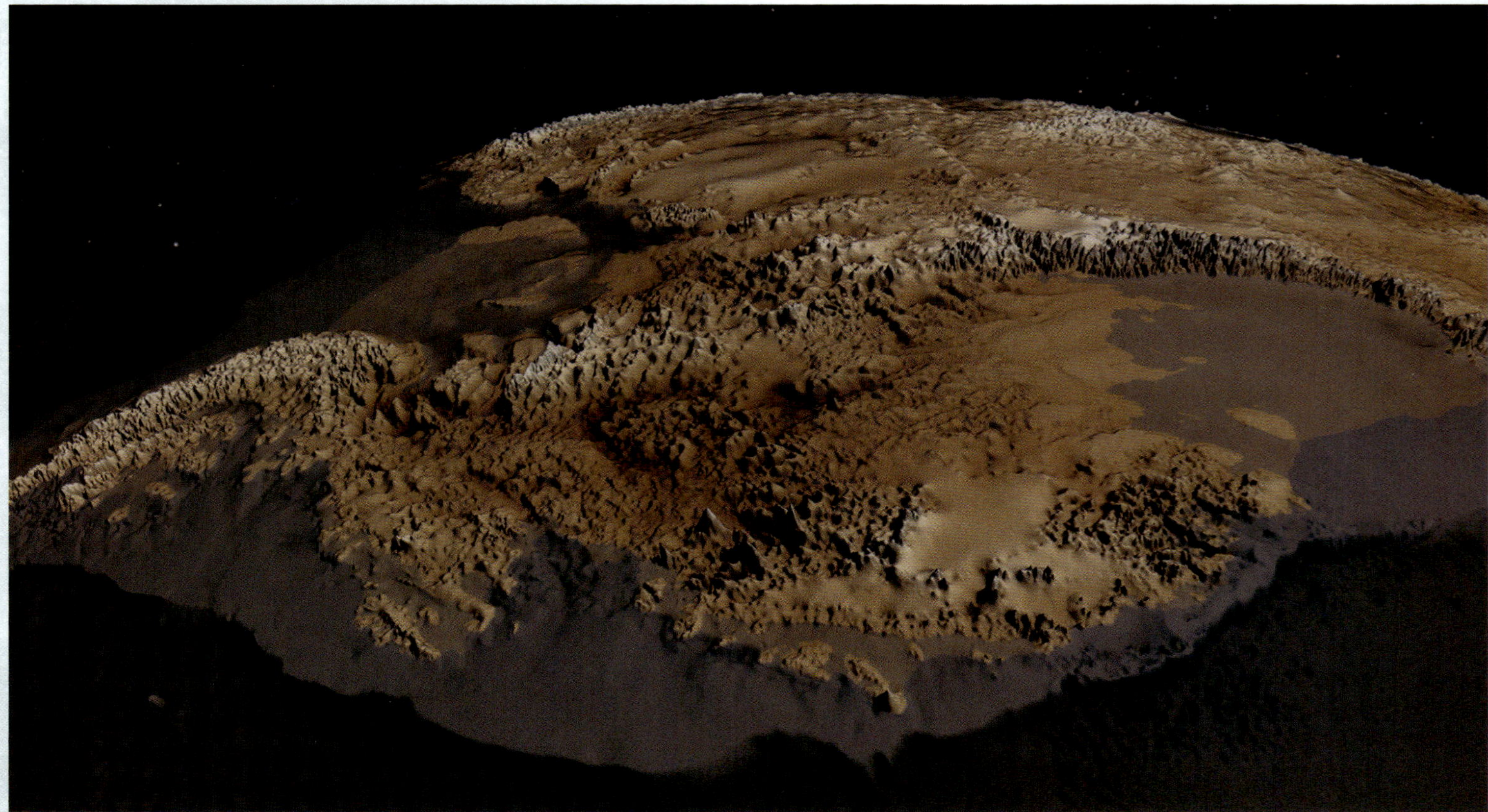

Figure 15.3. The East Antarctic plateau is shown at the top of the image. The Gamburtsev Mountains are near the center of this plateau. Elevations have been exaggerated for clarity.

East Antarctica lie beneath the ice. Ground-penetrating radar has revealed detailed information about the topography of these mountains. Most secular scientists think the mountains are nearly a billion years old, so they were expecting eons of erosion to have worn down and smoothed the mountain chain's rough edges. So scientists were stunned when the radar images revealed sharp, craggy features generally associated with young mountain ranges.[15] The Gamburtsev Mountains are located in Eastern Antarctica, in the upper right of Figure 15.3. Robin Bell, a geophysicist and expert on this part of Antarctica, explained the reason for the surprise:

> "It's really hard to imagine that there are mountains under there. It doesn't matter which way you spin—it's pretty flat," said Bell, who has studied the area for years. Yet, she added, the truly mysterious part of the hidden mountains is not *that* they exist, but how they *still* exist. The inexorable march of geological time erodes mountains away (if we came back in 100 million years, the Alps would be gone, Bell said) and the Gamburtsevs, at the ripe old age of 900 million to a billion years old, should have been worn down eons ago.[16]

"Yet, she added, the truly mysterious part of the hidden mountains is not *that* they exist, but how they *still* exist. The inexorable march of geological time erodes mountains away."

Of course, this lack of erosion is evidence that the mountains are quite young, as would be expected in the biblical worldview. Nevertheless, uniformitarian scientists claimed to have solved this apparent mystery in November 2011.[17] Supposedly, a collision between continents a billion years ago pushed up the Gamburtsev Mountains, but the root of the mountains remained embedded in the crust. Geological time eroded away the mountains—but a few hundred million years ago, a pulling apart, or rifting, of the crust near the root somehow warmed the root, causing it become less dense so that it floated up, forming a rejuvenated, relatively young, mountain chain. However, as Robin Bell acknowledged, the details of this explanation are still fuzzy:

Zircon

"That's [how the change in density of the mountain root occurred] the biggest thing that has us scratching our heads," Bell said. "We don't know if the rifting added a little heat, added a little water—we know the rifting happened, and [the mountain range] popped up, but we're still working on the question of how you make that phase change," she said.[18]

Furthermore, radioisotope dating of zircon crystals in sediments led other scientists to conclude that the mountains could not have formed recently but had to be (by secular reckoning) at least 500 million years old.[19,20]

Apparently, this first explanation was not entirely satisfactory, because scientists proposed another hypothesis three years later.[21] Heat from the earth's interior can warm the ice near the bedrock, forming rivers and lakes. Due to the high pressures deep beneath the ice, it is possible for the water to sometimes flow uphill. Scientists claim that this meltwater flowed uphill and then refroze above the mountains, protecting them from erosion. However, this proposed mechanism does not explain how the mountains were protected from erosion *before* the ice sheet formed.

> It's possible, she [Columbia University geochemist Sidney Hemming] added, that the ice has helped keep the rocks in place, but that does not explain the lack of erosion from ancient periods when the ice was not there.[22]

Of course, the lack of erosion is to be expected if both the mountains and the ice sheets are young.

Undoubtedly, other arguments can be made for the youthfulness of the Antarctic and Greenland ice sheets. Creation scientists are confident that as we obtain more data, the case for the youthfulness of the Greenland and Antarctic ice sheets will only become stronger.

References

1. Zielinski, G. A. et al. 1996. A 110,000-Yr record of explosive volcanism from the GISP2 (Greenland) ice core. *Quaternary Research.* 45 (2): 109.
2. Gow, A. J. and T. Williamson. 1971. Volcanic Ash in the Antarctic ice sheet and its possible climatic implications. *Earth and Planetary Science Letters.* 13 (1): 210-218.
3. Fuji, Y. et al. 1999. Tephra layers in the Dome Fuji (Antarctica) deep ice core. *Annals of Glaciology.* 29: 126.
4. Ibid, 127.
5. POV: Bill Nye: Science Guy. PBS, aired April 18, 2018.
6. Hebert, J. Bill Nye, PBS Highlight Young-Earth Evidence. *Creation Science Update.* Posted on ICR.org April 27, 2018, accessed June 13, 2018. Emphasis in the original.
7. Hebert, J. 2018. Tephra and Inflated Ice Core Ages. *Journal of Creation.* 32 (3): 4-6.
8. Narcisi, B., J. R. Petit, and B. Delmonte. 2010. Extended East Antarctic ice-core tephrostratigraphy. *Quaternary Science Reviews.* 29: 25.
9. Ibid, 25.
10. Ibid.
11. Ibid.
12. Ford, A. B. et al. Antarctica: Continent. *Encyclopaedia Britannica.* Posted on britannica.com, accessed June 14, 2018.
13. Barrett, P. 2003. Cooling a Continent. *Nature.* 421 (6920): 221-223.
14. Hvidberg, C. S. 2000. When Greenland Ice Melts. *Nature.* 404 (6778): 551-552.
15. Fountain of Youth Underlies Antarctic Mountains. The Earth Institute. Columbia University press release. Posted on earth.columbia.edu November 19, 2014, accessed June 14, 2018.
16. Mustain, A. Antarctica's Biggest Mysteries: Secrets of a Frozen World. *LiveScience.* Posted on livescience.com December 14, 2011, accessed June 14, 2018. Emphasis in the original.
17. Amos, J. Gamburtsev 'ghost mountains mystery solved.' BBC. Posted on bbc.com November 17, 2011, accessed June 14, 2018.
18. Mustain, Antarctica's Biggest Mysteries: Secrets of a Frozen World.
19. Van de Flierdt, T. et al. 2008. Evidence against a young volcanic origin of the Gamburtsev Subglacial Mountains, Antarctica. *Geophysical Research Letters.* 35 (21): L21303.
20. Carroll, R. Mystery Deepens Over Unseen Antarctic "Alps." *National Geographic News.* Posted on nationalgeographic.com November 6, 2008, accessed June 14, 2018.
21. Antarctic "ghost mountains" preserved by ice sheet. British Antarctic Survey news release. Posted on bas.ac.uk November 19, 2014, accessed June 14, 2018.
22. Carroll, Mystery Deepens Over Unseen Antarctic "Alps."

> vineyard. What grows of its own accord of your harvest you shall not reap, nor gather the grapes of your untended vine, for it is a year of rest for the land." (Leviticus 25:3-5)

Obviously, allowing the land to rest would allow the replenishment of nutrients in the soil depleted during the previous six years. The Israelites disobeyed this command, and during their captivity God caused the land to lie fallow 70 years, one year for each of the Sabbath years this command was disobeyed (2 Chronicles 36:20-21). God also commanded the Israelites to refrain from cutting down fruit trees during the siege of an enemy city.

> "When you besiege a city for a long time, while making war against it to take it, you shall not destroy its trees by wielding an ax against them; if you can eat of them, do not cut them down to use in the siege, for the tree of the field is man's food. Only the trees which you know are not trees for food you may destroy and cut down, to build siegeworks against the city that makes war with you, until it is subdued." (Deuteronomy 20:19-20)

Even here it is obvious that this command is primarily for humanity's benefit. Likewise, the Bible makes it clear that God expects people to treat the animals under their dominion with kindness.

> A righteous man regards the life of his animal, but the tender mercies of the wicked are cruel. (Proverbs 12:10)

A righteous man regards the life of his animal. Proverbs 12:10

The expectation that the righteous will treat their animals humanely does not negate the fact that God has given people dominion over animals (Genesis 1:28). Nor does it negate the fact that God has given us permission to eat animal meat (Genesis 9:3; 1 Timothy 4:3). Scripture itself makes it clear that one day God will judge those who destroy the earth:

> The nations were angry, and Your wrath has come, and the time of the dead, that they should be judged, and that You should reward Your servants the prophets and the saints, and those who fear Your name, small

> and great, and should destroy those who destroy the earth. (Revelation 11:18)

However, the biblical material addressing this subject is a relatively small percentage of the total biblical text. The Bible often discusses humanity's general accountability to God (Ecclesiastes 12:13-14; Matthew 25:14-30; 2 Corinthians 5:10; Hebrews 9:27; 1 Peter 4:17), but it does not devote much time or space to the subject of humans' stewardship of the earth. This is not too surprising given that the focus of Scripture is on people's relationship with God and their need for salvation from death and the judgment to come. In light of this weighty truth, it is hardly surprising that the Bible spends relatively little time addressing the subject of conservation.

Environmentalism and God's Providence

At the root of climate alarmism is a denial of God's providence for His creatures. In the biblical worldview, God has designed this planet as our home and as the future residence of the King of kings. Since an all-knowing, all-powerful, and loving Creator made us and our home, we can reasonably expect that He designed our home with robust climate feedbacks to prevent or minimize climate extremes. Recall that experts have testified that climate stability is the heart of the debate over global warming.[4] The Christian worldview provides us with a theological rationale for suspecting that our climate system is stable. In the next chapter, we will explore scientific reasons for this conclusion.

Of course, some would object that it is dangerous and reckless to allow religion to influence our understanding of climate or proposed climate policy. However, it is impossible to avoid the subject of origins when interpreting past climate data. The secularist may fancy that he is free from such influence, but he is not. Everyone, even an atheist, has a religious worldview that influences the way he looks at the world. I have attempted to show in previous chapters that the Christian worldview makes much better sense of the scientific and historical data than does the evolutionary worldview, both in general terms and in regard to past climate change. If that is so, then it is completely reasonable to allow the Christian worldview to affect the way that one looks at the world.

In contrast, it is secularists who are actually engaging in reckless behavior. Their naturalistic origins myth is completely untenable and without scientific support. They

Torr Head, Northern Ireland

cannot hope to explain the origin of life and have, in moments of candor, acknowledged as much.[5] They cannot even plausibly explain the Ice Age, the most recent major geological event in Earth history.[6] Yet, they are willing to use this nontenable worldview to draw conclusions about future climate change. So, it isn't really possible to keep religion out of the climate change debate. What we believe about the past will influence what we believe about the future.

References

1. Morris, H. M. 1993. *The Genesis Record: A Scientific and Devotional Commentary on the Book of Beginnings*. Grand Rapids, MI: Baker House Books, 46-48.
2. Morris, H. M. 1997. Why the Gap Theory Won't Work. *Acts & Facts* 26 (11).
3. Morris, *The Genesis Record,* 76-77.
4. Curry, J. Lewis and Curry: Climate Sensitivity Uncertainty. Posted on judithcurry.com September 24, 2014, accessed January 4, 2017.
5. Horgan, J. Pssst! Don't tell the creationists, but scientists don't have a clue how life began. *Scientific American.* Posted on scientificamerican.com February 28, 2011, accessed December 17, 2018.
6. Oard, M. J. 1990. *An Ice Age Caused by the Genesis Flood*. El Cajon, CA: Institute for Creation Research, 196.

17 Why Uniformitarian Climate Interpretations of Chemical Wiggles Are Doubtful

Summary: Climate researchers pull sediment cores from the ocean floor. They examine the oxygen content of shells of tiny microorganisms called foraminifera to make determinations about past climate change. However, this attempt to infer clues about past climate change faces many practical difficulties. Furthermore, the absence of metallic pellets called manganese nodules in the deeper seafloor sediments is a clue that ocean sediments were once deposited much more rapidly than they are today, contradicting "slow and gradual" secular expectations. Because manganese nodules stop growing when covered by more than a few centimeters of sediment, creation scientists think that sediments toward the end of the Flood and afterward were deposited so rapidly that nascent manganese nodules did not have time to grow to any appreciable size. This extremely rapid sedimentation was likely caused by two factors. Receding floodwaters scoured enormous amounts of sediment off the continents, forming large planation surfaces on every continent and quickly dumping the sediments onto the continental margins. Likewise, oceans during and after the Flood were quite warm and mineral-rich, which greatly increased phytoplankton, a food source for foraminifera, also called forams. This in turn greatly increased the number of forams and the rate at which their remains contributed to the accumulating seafloor sediment. The failure to take these factors into account is a major source of error in uniformitarian interpretations of the seafloor sediment data.

Secular researchers of past climates (called *paleoclimatologists*) think chemical wiggles in foraminifera (foram) shells are telling a story about past climate change. Understanding how this is supposed to work will help shed some light on why the interpretation of these chemical wiggles is tricky.

Oxygen Isotope Values in Seafloor Sediments

Remember that some oxygen atoms are heavier than other oxygen atoms. Specifically, the ^{18}O variety or isotope of the oxygen atom is heavier than the ^{16}O variety. This means that water (H_2O) molecules containing an ^{18}O atom rather than a ^{16}O atom will be a little heavier. Likewise, H_2O molecules with a ^{16}O atom will be lighter.

It is easier for the lighter H_2O molecules to evaporate from the surface of the ocean than for heavier H_2O molecules. This means that at a given temperature, one expects the ocean would be slightly depleted in these lighter water molecules (those containing ^{16}O atoms), but the water vapor in the air would be slightly enriched with them. Furthermore, this effect is greater at lower temperatures.

This effect is also seen in foraminifera shells. The forams use both the heavier ^{18}O atoms and the lighter ^{16}O atoms to construct their calcium carbonate ($CaCO_3$) shells, and the number of ^{18}O atoms compared to the number of ^{16}O atoms depends on temperature. So, could scientists use measurements of the number of ^{18}O and ^{16}O atoms in

Foraminifera

foram shells to calculate oxygen isotope ratio ($\delta^{18}O$) values for those foram shells? And could those foram $\delta^{18}O$ values be used to infer the temperature of the surrounding water at the time a particular foram made its shell?

In theory, yes. However, there is a catch. Not only does the foram $\delta^{18}O$ value depend on the temperature of the surrounding seawater, it also depends on the chemistry of that seawater, specifically the seawater's $\delta^{18}O$ value. Unfortunately, this value cannot be measured because this was the $\delta^{18}O$ value of the seawater at the time the foram was making its shell—*not* the present value. Scientists do not have time machines that allow them to go back in time to make this measurement.

For this reason, the precise meaning of these foram $\delta^{18}O$ values is not obvious. Today, secular scientists think they indicate past amounts of global ice volume, but they did not always think this.[1,2] They used to think that these values were primarily indicators of temperature.[3]

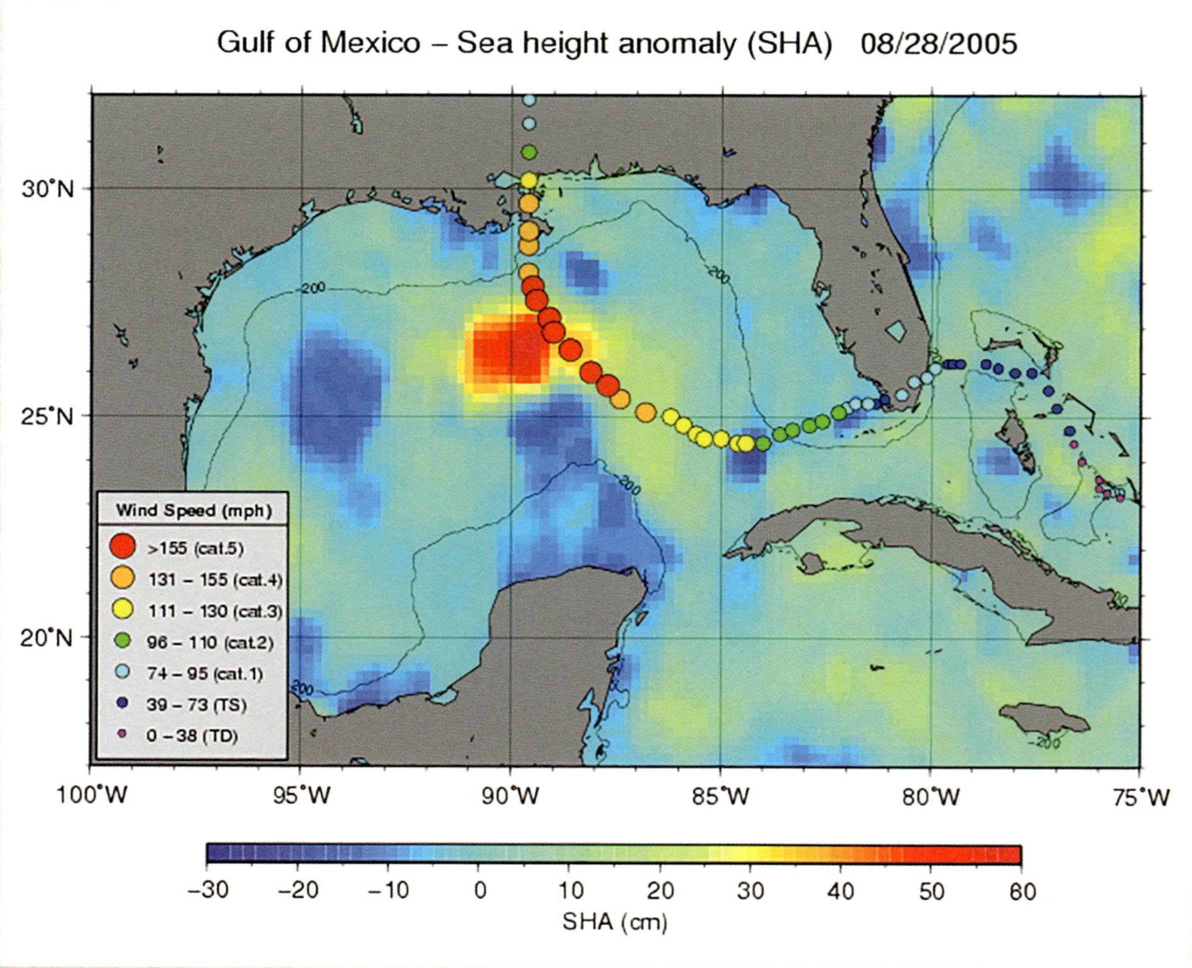

Figure 17.1. Hurricane Katrina passing over a warm core ring in the Gulf of Mexico

Furthermore, local factors can affect the $\delta^{18}O$ values within foraminifera shells, especially the free-floating kind known as *planktonic foraminifera.* Ocean temperatures vary dramatically with depth in a narrow vertical zone called the *thermocline.* This means that $\delta^{18}O$ values in foram shells can be influenced by temperature changes resulting from changes in the depths at which the foram was floating as it constructed its shell. Likewise, temperature can vary significantly in the horizontal direction due to large rotating eddies called warm and cold cores (Figure 17.1). These eddies can persist for long periods of time. Remember that the $\delta^{18}O$ values are supposed to be *global* climate indicators, but there is no way scientists can know how these unknown local factors influenced $\delta^{18}O$ values as the forams were making their shells. Creation researcher Michael Oard pointed out these difficulties more than 30 years ago.[4]

Worse yet, scientists recently discovered another problem. After a free-floating (or planktonic) foram constructs its shell, the surrounding environment can subtly alter the shell's chemistry. That in turn alters the apparent average $\delta^{18}O$ value for the shell. Researchers have known about this issue for a long time, so they generally try to use only pristine foram shells (shells showing no obvious changes due to the environment) in their research. However, they recently realized that the chemistry of even pristine-looking shells can be altered sufficiently to greatly skew the interpretation of the $\delta^{18}O$ values.[5,6]

The Creation Interpretation

People often ask me how creation scientists interpret these oxygen isotope wiggles. Creationists agree that, generally speaking, higher $\delta^{18}O$ values in seafloor sediments likely correspond to colder temperatures, and lower $\delta^{18}O$ values correspond to warmer temperatures. However, since all the seafloor sediments were deposited during and after the Genesis Flood, and since only one Ice Age followed the Flood, these seafloor chemical wiggles can't possibly represent global ice volume, as claimed by uniformitarian scientists. Furthermore, the factors described above complicate the interpretation of these chemical wiggles, making a simple one-size-fits-all interpretation very difficult, if not impossible. Dr. Larry Vardiman suggested that cycles in the seafloor sediment $\delta^{18}O$ values might represent cyclic events that are still occurring in the present-day, such as the El Niño/La Niña weather oscillation.[7]

However, one thing is certain. Before you can even begin to determine the cause

> "However, since all the seafloor sediments were deposited during and after the Genesis Flood, and since only one Ice Age followed the Flood, these seafloor chemical wiggles can't possibly represent global ice volume, as claimed by uniformitarian scientists."

of these $\delta^{18}O$ wiggles, you first need an accurate timescale for the deep-sea sediment cores. Obviously, creation scientists think that the uniformitarian ages for these cores are grossly in error. One reason for this is given below.

Manganese Nodules: Evidence for Rapid Sediment Deposition

Creation scientists think the distribution of manganese nodules within the deep-sea sediments presents a strong argument that most seafloor sediments accumulated rapidly on the ocean floor, not slowly and gradually as secular scientists claim.[8] Manganese nodules are metallic pellets found on the ocean floor (Figure 17.2). These pellets are usually composed of manganese but can also be made of nickel, iron, copper, and other metals. Scientists think these nodules form when chemicals in seawater precipitate onto a small object that can serve as a nucleus for growth, such as a shark's tooth. Based upon radioisotope dating methods, uniformitarians insist these nodules grow in size extremely slowly, just a few millimeters per million years.

Yet, researchers have repeatedly observed these nodules growing rapidly in man-made reservoirs and lakes. These nodules are sometimes located on underwater debris and wreckage from World Wars I and II. The nodules' observed growth rates are about a hundred thousand times faster than the rates calculated from radioisotope dating! Obviously, we creationists think this is just one more indication that there is something seriously wrong with radioisotope dating assumptions.

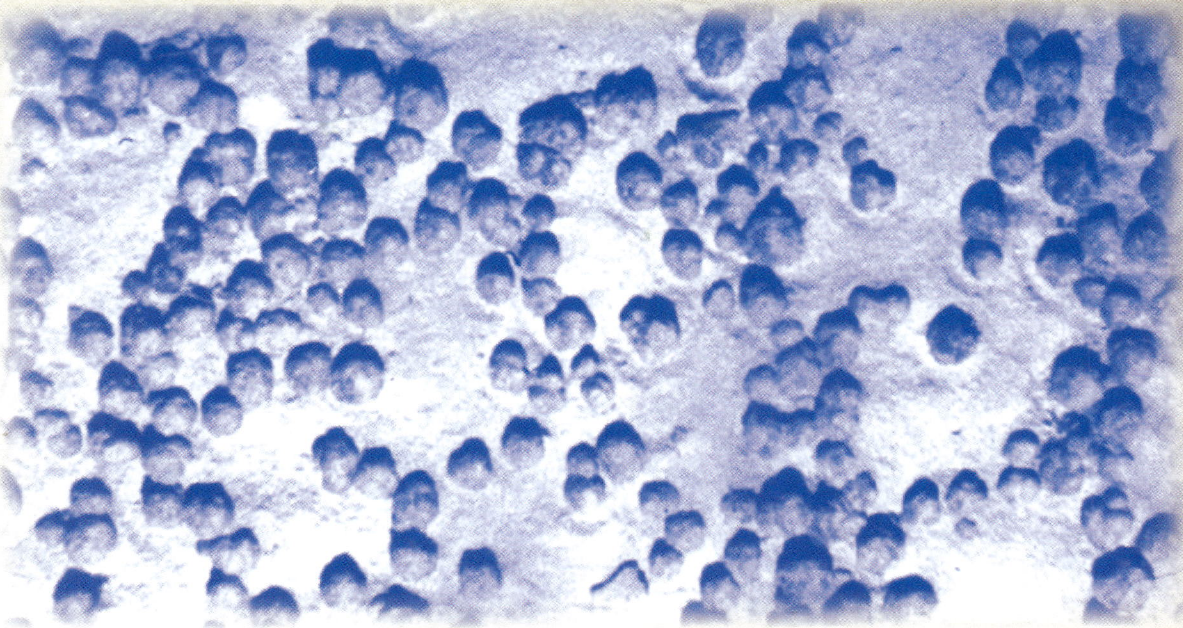

Figure 17.2. Manganese nodules on the ocean floor

Two important facts about these nodules require explanation. First, scientists think nodules stop growing once they are covered by a few centimeters of sediments.[9,10] Second, most nodules are located only within the uppermost few meters of sediments, with the very highest density of nodules in the uppermost 50 centimeters.[8,11] Scientists sometimes find nodules at greater depths, but these deeper nodules tend to be smaller than nodules near the surface. Remember that deep-sea cores are the only means that scientists have for inferring how these nodules are distributed with depth in the sediments. Some researchers think it's likely that some of the deeper nodules fell down drill holes during the coring process.

Secular scientists acknowledge that this general absence of nodules in the deeper seafloor sediments does not appear to be because the nodules are dissolving. Some have suggested that bottom ocean currents or bottom-dwelling ocean creatures disturbed the sediments so that these uppermost nodules are exposed to seawater for longer times than the more deeply buried nodules. This greater exposure time allows these uppermost nodules to grow larger. This is a logical explanation, but it violates the uniformitarian assumption that "the present is the key to the past." Why would these nodules grow to larger sizes *only* in the relatively recent past? If ocean currents or bottom-dwelling organisms could disturb sediments in the recent past, then why could they also not disturb them in the distant past, too?

Creation scientists have an alternate explanation (Figure 17.3). Nodules are generally absent from the deeper seafloor sediments because those sediments accumulated so rapidly that nascent nodules simply had no opportunity to grow to any appreciable size.[8,12] Toward the end of the Genesis Flood, receding floodwaters rapidly eroded enormous amounts of material from the continents and deposited those sediments into the ocean basins.

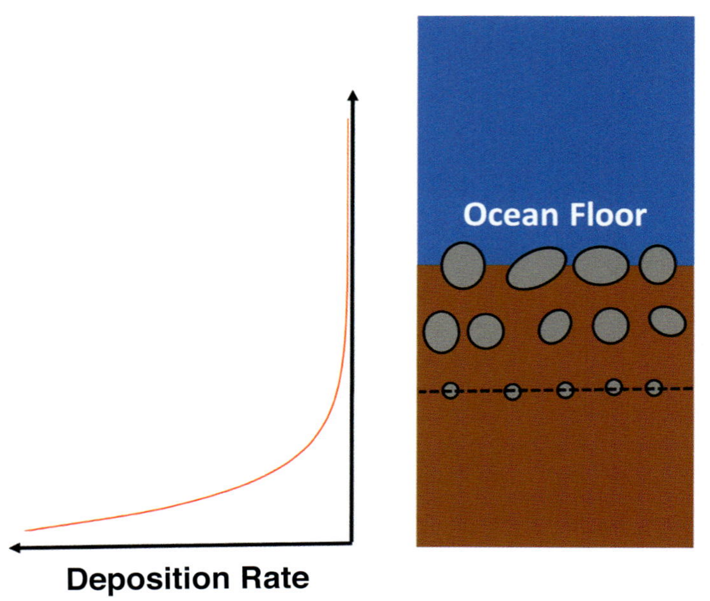

Figure 17.3. Both the decreasing size of manganese nodules with depth in ocean sediments and their absence from much deeper sediments can be explained if most seafloor sediments were deposited too rapidly for nascent nodules to grow to any appreciable size. One would expect sedimentation rates to be very high at the end of the Genesis Flood and shortly afterward, but these high rates would gradually decrease, allowing manganese nodules to form in the uppermost seafloor sediments.

Plenty of evidence can be found for rapid erosion. Every continent contains large flat areas called *planation surfaces*. These are not forming today, and secular scientists have a difficult time explaining their existence. Some planation surfaces show evidence that very hard rocks were eroded just as much as the softer rocks or sediments. This is consistent with catastrophically rapid erosion, like the erosion that would have occurred toward the end of the Flood. In that case, fast-moving sheets of water indiscriminately planed down both the harder and softer areas.[13,14]

Warm, mineral-rich oceans during and after the Flood would have stimulated phytoplankton growth. Since foraminifera feed on phytoplankton, foraminifera also greatly increased in number after the Flood. The remains of these abundant foraminifera also contributed to the quickly accumulating sediments.[15] If most of the sediments accumulated rapidly and catastrophically just thousands of years ago, this invalidates any interpretation of the chemical wiggles that assumes the sediments accumulated slowly over hundreds of thousands of years.

Chemical Wiggles in Ice Cores

Secular scientists also make $\delta^{18}O$ measurements within ice cores. The interpretation of these oxygen isotope values in ice cores is a little different from that of the $\delta^{18}O$ values in the seafloor sediments. In the ice cores, low $\delta^{18}O$ values represent Ice Ages, and high $\delta^{18}O$ values represent warmer periods, the reverse of what is true in the seafloor sediment cores.[16]

But there are factors that complicate the interpretation of these chemical wiggles in the ice. Oxygen isotope values in seawater also depend upon latitude. This means that ice $\delta^{18}O$ values are also influenced by the distance traveled by the cloud that deposited the snow or ice. Remember also that scientists are only measuring $\delta^{18}O$ values at places where they have drilled ice cores. But suppose that during an Ice Age an ^{18}O-depleted cloud deposited some rain or snow before reaching the ice core location. Because it is easier for the heavier H_2O molecules to precipitate, the cloud will be even more depleted in ^{18}O than it would have been had it only deposited snow at the ice core's location. This means that $\delta^{18}O$ variations in the ice can be influenced by how much precipitation was *previously* deposited by the cloud before it deposited the snow that would eventually become part of the ice core.

But there is absolutely no way that any scientist—creationist or evolutionist—could

Cape Breton Highlands planation surface in Nova Scotia, Canada

possibly know how far a cloud that existed thousands of years ago traveled before depositing snow at a particular location, nor is there any way to know whether that cloud did or did not deposit rain or snow at some other location before doing so. So variations in ice core $\delta^{18}O$ values (or other such chemical wiggles) that are generally attributed to global climate changes were likely influenced by other factors, a possibility acknowledged even by secular scientists.[17,18]

However, creationists would agree that lower (more negative) $\delta^{18}O$ values in ice cores generally correspond to colder temperatures, and higher (less negative) $\delta^{18}O$ values correspond to warmer temperatures. For instance, Michael Oard thinks that many of the ice core $\delta^{18}O$ variations represent decades-long temperature changes due to varying amounts of volcanic aerosols. Even so, the interpretation of these chemical wiggles is definitely not as simple as some might suppose.

References

1. Shackleton, N. 1967. Oxygen Isotope Analyses and Pleistocene Temperatures Re-assessed. *Nature.* 215: 15-17.
2. Wright, J. D. 2010. Cenozoic Climate—Oxygen Isotope Evidence. In *Climates and Oceans.* J. H. Steele, ed. Amsterdam, Netherlands: Academic Press, 316-327.
3. Emiliani, C. 1966. Isotopic Paleotemperatures. *Science.* 154 (3751): 851-857.
4. Oard, M. J. 1984. Ice Ages: The Mystery Solved? Part II: The Manipulation of Deep-Sea Cores. *Creation Research Society Quarterly* 21 (3): 125-137.
5. Oard, M. J. 2003. The 'cool-tropics paradox' in palaeoclimatology. *Journal of Creation.* 17 (1): 6-8.
6. Wycech, J., D. C. Kelly, and S. Marcott. 2016. Effects of seafloor diagenesis on planktic foraminiferal radiocarbon ages. *Geology.* 44 (7): 551-554.
7. Vardiman, L. 2001. *Climates Before and After the Genesis Flood: Numerical Models and Their Implications.* El Cajon, CA: Institute for Creation Research, 79-80.
8. Patrick, K. 2010. Manganese nodules and the age of the ocean floor. *Journal of Creation.* 24 (3): 82-86.
9. Glasby, G. P. 1978. Deep-sea manganese nodules in the stratigraphic record: evidence from DSDP cores. *Marine Geology.* 28 (1-2): 51-64.
10. Pattan, J. N. and G. Parthiban. 2007. Do manganese nodules grow or dissolve after burial? Results from the Central Indian Ocean Basin. *Journal of Asian Earth Sciences.* 30 (5-6): 696-705.
11. Somayajulu, B. L. K. 2000. Growth rates of oceanic manganese nodules: Implications to their genesis, palaeo-earth environment and resource potential. *Current Science.* 78 (3): 300-308.
12. Hebert, J. Manganese Nodule Discovery Points to Genesis Flood. *Creation Science Update.* Posted on ICR.org March 5, 2015, accessed November 27, 2018.
13. Oard, M. J. 2006. It's plain to see: Flat surfaces are strong evidence for the Genesis Flood. *Creation.* 28 (2): 34-37.
14. Hebert, J. and T. Clarey. 2015. Ice Cores, Seafloor Sediments, and the Age of the Earth, Part 3. *Acts & Facts.* 44 (1): 10-13.
15. Vardiman, L. 1997. Global Warming and the Flood. *Acts & Facts.* 26 (12).
16. A discussion of why this is the case can be found in most paleoclimatology textbooks.
17. Petit, J. R. et al. 1991. Deuterium excess in recent Antarctic snow. *Journal of Geophysical Research.* 96: 5113-5122.
18. Vardiman, *Climates Before and After the Genesis Flood,* 55-68.

18 Scientific Evidence for Low Climate Sensitivity

Summary: The main issue in the climate change controversy is whether Earth's climate is stable or unstable—whether it is sensitive or insensitive to changes in atmospheric carbon dioxide. A major argument for high climate sensitivity is that computer models predict dangerous future warming. However, in retrospect past model predictions of temperature increases were almost always too high. A big problem with computer models is that there is still much about climate that is poorly understood, and modellers have to guess how these factors will influence the result. Scientists on both sides of the debate acknowledge that clouds are a major source of uncertainty. A cloud can either have a cooling or warming effect depending upon the size of its droplets, its altitude, and its latitude. However, scientists do not understand very well the processes within clouds that can influence droplet sizes. So, computer climate models are a poor argument for high climate sensitivity. On the other hand, there is positive scientific evidence that our climate is stable and self-regulating. Vegetation is one of these regulating mechanisms. Climate alarmists are concerned about the rise of carbon dioxide, but what consumes carbon dioxide? Plants! As carbon dioxide increases, so does plant growth, offering a natural method of regulation.

As noted in chapter 3, experts have acknowledged that climate sensitivity is the real issue in the global warming/climate change controversy.[1,2] A doubling of atmospheric carbon dioxide will result in a small increase in global average temperatures—assuming that everything else stays the same. However, the climate system is complicated, and a doubling of atmospheric carbon dioxide will change other aspects of the climate system. These other changes could either enhance or impede this warming. Those who are most concerned about the issue of climate change think that our climate system is very sensitive to changes. Why do they think this?

One reason is that climate computer models predict significant warming as a result of increased atmospheric CO_2. To put it another way, most climate computer models suggest that the equilibrium climate sensitivity is high (Figure 18.1). However, past temperature predictions of these computer models have often been too high (Figure 18.2).[3] Hence, there is good reason to think that the computer models are flawed.

"However, past temperature predictions of these computer models have often been too high. Hence, there is good reason to think that the computer models are flawed."

Equilibrium climate sensitivity is a quantity used by scientists to indicate the sensitivity of Earth's climate to increases in some parameter such as atmospheric

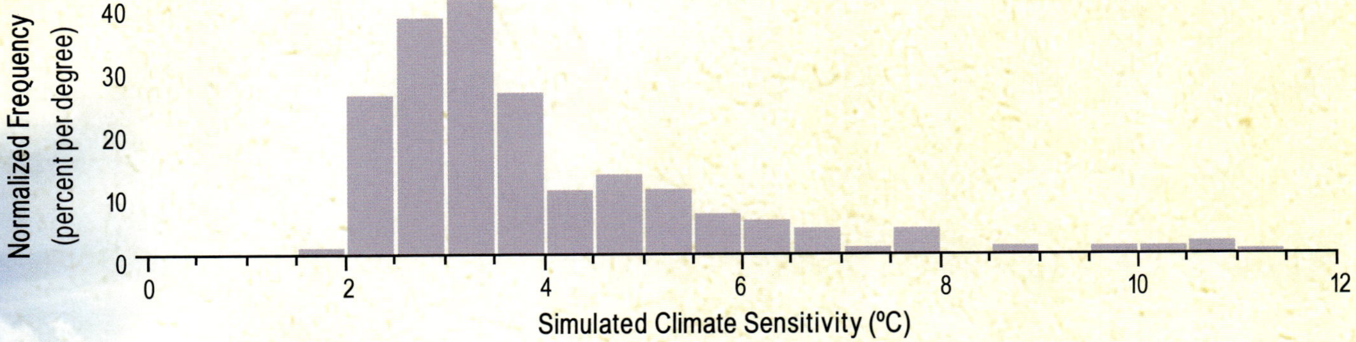

Figure 18.1. The results of most computer climate simulations suggest high climate sensitivity, with a resulting temperature increase of at least 2°C in response to a doubling of atmospheric carbon dioxide, and perhaps greater than 10°C

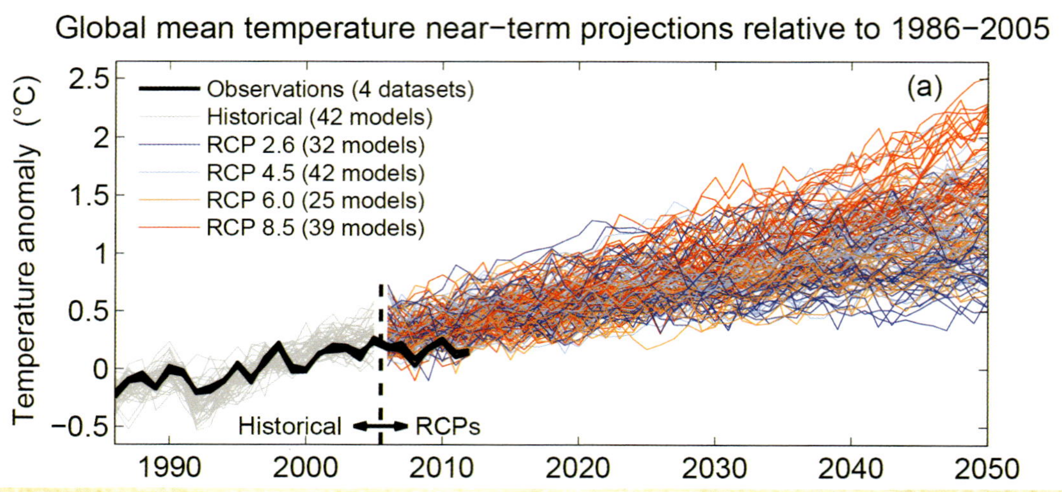

Figure 18.2. Climate computer models (thin gray and colored lines) have consistently over-predicted the amounts of observed past warming (thick black line)[3]

carbon dioxide. Here we denote this quantity by the Greek letter λ. This quantity is defined to be the equilibrium change in Earth's global average surface temperature if the amount of atmospheric carbon dioxide doubles. It is called an *equilibrium* climate sensitivity because it indicates the change in global average surface temperature after the climate system has settled down.

Computer climate models tend to yield estimates of λ that are on the high side, which is one reason that many are concerned about this issue. However, NASA climatologist and meteorologist Dr. Roy Spencer had this to say about these estimates on his personal blog:

> Remember, the sensitivity of their [computer] models is NOT the result of basic physics, as some folks claim...it's the result of very uncertain parameterizations (e.g. clouds) and assumptions (e.g. precipitation efficiency effects on the atmospheric water vapor profile and thus feedback). The models are adjusted to produce warming estimates that "look about right" to the modelers. Yes, *some* amount of warming from increasing CO_2 is reasonable from basic physics. But just *how much* warming is open to manipulation within the uncertain portions of the models.[4]

Dr. Spencer was writing in response to an important paper by climatologist Judith Curry (formerly of Georgia Institute of Technology[5]) and independent researcher

"The models are adjusted to produce warming estimates that 'look about right' to the modelers."

Clouds are one of the largest uncertainties in climate computer models

Nicholas Lewis.[6,7] This paper used historical climate data and the principle of conservation of energy to obtain a more accurate assessment of λ. They concluded that λ had a value of about 1.66°C, which was toward the low end of the range of estimates that are usually given for this value (between 1.5°C and 4.5°C). In other words, they estimated that climate sensitivity is relatively low.

Such a result is consistent with previous evidence that our climate system "self-regulates" to avoid extremes. For instance, some desert areas are shrinking in size as the amount of surrounding vegetation increases. Moreover, this increase in vegetation seems to be a response to increased atmospheric CO_2.[8] Likewise, it seems reasonable that algae could help to naturally balance carbon dioxide levels.[9,10] The model-generated evidence for an unstable climate is weak, at best.[11]

"Such a result is consistent with previous evidence that our climate system 'self-regulates' to avoid extremes."

As an aside, Dr. Spencer is absolutely correct that clouds are a major source of uncertainty in computer climate models. This is acknowledged by both sides of the debate:

> Clouds and aerosols continue to contribute the largest uncertainty to estimates and interpretations of the Earth's changing energy budget.[12]

So, a better understanding of clouds would help shed light on this issue. Interestingly, my Ph.D. advisor, Dr. Brian Tinsley, professor emeritus at the University of Texas at Dallas, has studied this issue for almost 30 years and concluded that current models are likely not taking into account all the relevant physics. In fact, as hard as it may be for some to believe, I think it is highly likely that Dr. Tinsley has found the long-sought connection between solar activity, cosmic rays, and weather and climate. In fact, his work is briefly mentioned in the IPCC Fifth Assessment Report, although the authors are much too quick to dismiss it, in my opinon.[13] For interested readers, I discuss Dr. Tinsley's working hypothesis in a series of four open-access technical papers (and yes, they *are* technical). For interested readers, open links to these four papers are provided at ICR.org/jake_hebert. I also discuss his idea in more detail in

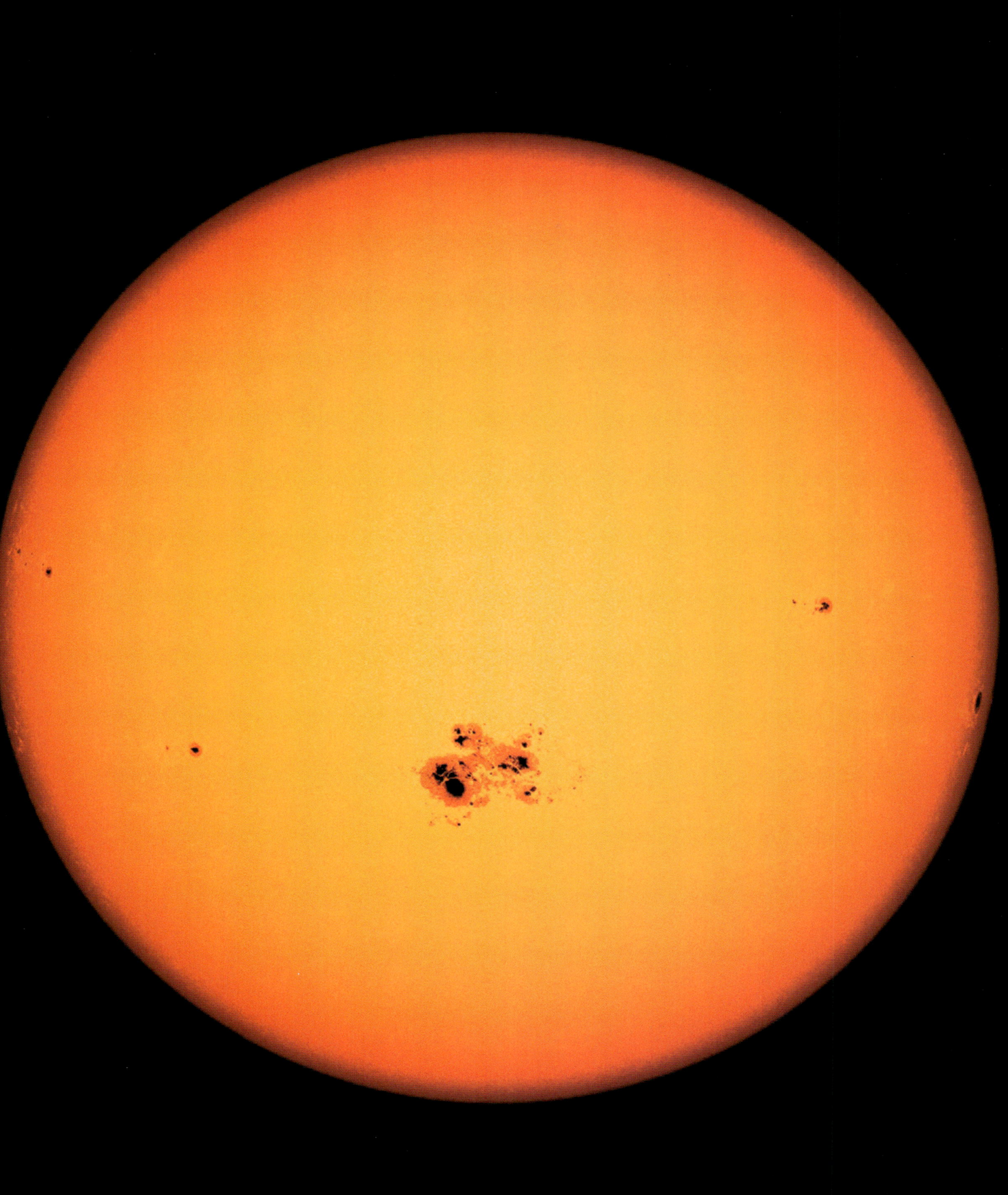

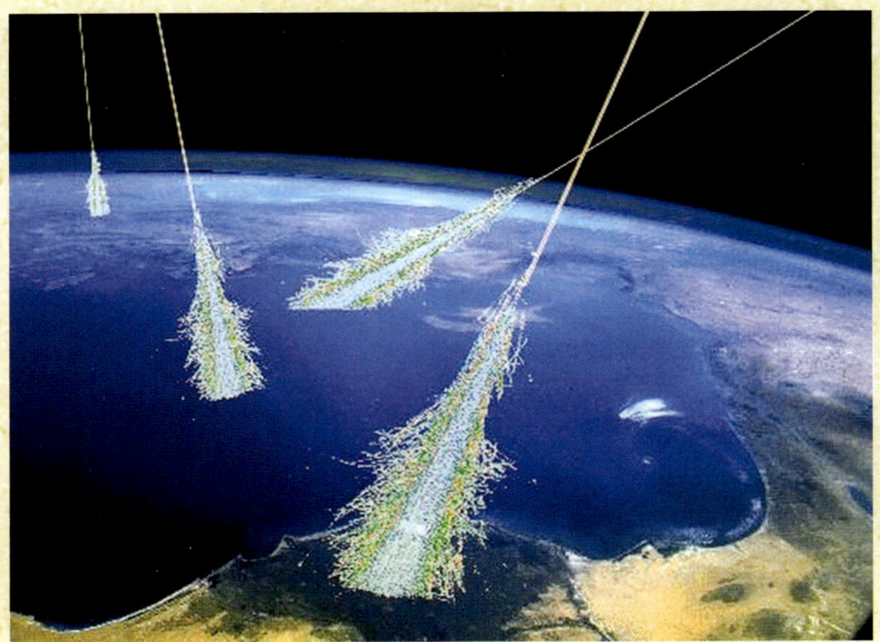

Cosmic rays produce a shower of charged particles when they enter the atmosphere

appendix A of this book. For the record, it should be noted that Dr. Tinsley thinks man-made global warming is real (although he does not think it is cause for panic), and he does not agree with my creationist views.

When cosmic rays enter the atmosphere, a shower of charged particles is produced. Scientists have long speculated that the sun could be influencing the number of cosmic rays entering the atmosphere, thereby influencing cloud properties and affect-

"Scientists have long speculated that the sun could be influencing the number of cosmic rays entering the atmosphere, thereby influencing cloud properties and affecting weather and climate."

ing weather and climate. More cosmic rays enter the atmosphere when the number of sunspots is low.

References

1. Spencer, R. Global Warming 101. Posted on drroyspencer.com, accessed November 8, 2016.
2. Curry, J. Lewis and Curry: Climate Sensitivity Uncertainty. Posted on judithcurry.com September 24, 2014, accessed November 2, 2014.
3. Figure TS.14 in Stocker, T. F. et al. 2013. *Climate Change 2013: The Physical Science Basis*. T. F. Stocker et al, eds. New York: Cambridge University Press, 87.
4. Spencer, R. New Lewis & Curry Study Concludes Climate Sensitivity is Low. Posted on drroyspencer.com April 24, 2018, accessed October 8, 2018. Emphasis in the original.
5. Hebert, J. Well-Known Scientist Resigns, Cites Climate Craziness. *Creation Science Update*. Posted on ICR.org January 19, 2017, accessed October 8, 2018.
6. Lewis, N. and J. Curry. 2018. The Impact of Recent Forcing and Ocean Heat Uptake Data on Estimates of Climate Sensitivity. American Meteorological Society.
7. Nic Lewis pointed out a significant error in a 2018 *Nature* paper that purported to show that the oceans were warming much faster than predicted, a mistake publicly acknowledged by the paper's authors. See his editorial, posted on judithcurry.com November 6, 2018, accessed November 15, 2018. See also Resplandy, L. et al. 2018. Quantification of ocean heat uptake from changes in atmospheric O_2 and CO_2 composition. *Nature*. 563: 105-108.
8. Thomas, B. Global Warming? Trees to the Rescue! *Creation Science Update*. Posted on ICR.org July 22, 2013, accessed October 8, 2018.
9. Thomas, B. Does Earth Balance Carbon Dioxide Levels Automatically? *Creation Science Update*. Posted on ICR.org January 12, 2009, accessed October 8, 2018.
10. Vardiman, L. 1997. Global Warming and the Flood. *Acts & Facts*. 26 (12).
11. We would expect our climate system to self-regulate if our world was designed by an all-knowing, all-powerful Creator who is concerned with our well-being. Thomas, B. Well-Engineered Ecosystems Bounce Back. *Creation Science Update*. Posted on ICR.org June 11, 2009, accessed October 8, 2018.
12. Boucher, O. et al. 2013. Clouds and Aerosols. In *Climate Change 2013: the Physical Science Basis. Contribution of Working Group I to the Fifth Assessment Report of the Intergovernmental Panel on Climate Change*. T. F. Stocker et al., eds. Cambridge and New York: Cambridge University Press: 573. Posted on ipcc.ch, accessed May 5, 2019.
13. Boucher, Clouds and Aerosols, 614.

19 How Uniformitarianism Misleads Paleoclimate Researchers

Summary: The Milankovitch theory is the source of most global warming alarmism. This theory is leading uniformitarian scientists to conclude that Earth's climate is unstable because the changes in sunlight predicted by the Milankovitch theory are too small to themselves be the sole cause of Ice Ages. Therefore, secular scientists concluded that other factors, including atmospheric carbon dioxide, are amplifying these small changes in sunlight to bring about major climate change. Scientists use the Milankovitch theory and orbital tuning to assign age estimates to seafloor sediments, and these incorrect age estimates are contributing to needless concern about rapidly rising sea levels. The failure of uniformitarian climatologists to recognize the unique, never-to-be-repeated nature of the Ice Age is also causing them to draw false analogies between our present-day climate and supposed past interglacials between repeating Ice Ages.

I have spent much of the previous 18 chapters presenting the case that the creation/Flood model does a much better job of explaining the geological and paleoclimate data than does uniformitarian thinking. The rocks and fossils are the result of the Genesis Flood 4,500 years ago, not the result of slow, gradual deposition over millions of years. So, any attempt to infer past changes in climate that assumes an old earth and uniformitarianism will be in error. This will be true in general, but here I would like to examine some specific ways that uniformitarianism misleads paleoclimate researchers.

An Unstable Climate?

A uniformitarian interpretation of paleoclimate data is contributing to climate change alarmism. Specifically, the Milankovitch Ice Age theory is leading uniformitarian scientists to conclude that our climate is unstable. Why is this?

As discussed earlier, the Milankovitch theory claims that very slow changes over tens of thousands of years in Earth's orbital and rotational motions alter the seasonal and latitudinal distribution of sunlight falling on the earth. These changes supposedly control the timing of Ice Ages. Uniformitarian scientists have concluded that the Milankovitch theory is correct (see chapter 9), but they also recognize that these changes in sunlight distribution are too small to be a sufficient cause of big climate change. For this reason, they are forced to conclude that "positive feedbacks" in the climate system can amplify the effect of these small changes to bring about an Ice Age. In other words, the Milankovitch Ice Age theory requires a high climate sensitivity, in spite of the empirical evidence against high climate sensitivity (see the previous chapter). Atmospheric scientist and emeritus ICR researcher Larry Vardiman noted:

> A major result of this need for feedback mechanisms has been the development of a perspective that the earth's climate system is extremely sensitive to minor disturbances. A relatively minor perturbation could initiate a non-linear response which might lead to another "Ice Age" or

> "Greenhouse." Because of the fear that a small perturbation might lead to serious consequences, radical environmental policies on the release of smoke, chemicals, and other pollutants and the cutting of trees have been imposed by international agencies and some countries. If the basis for the Astronomical Theory is wrong, many of the more radical environmental efforts may be unjustified.[1]

If you do an internet search for "climate instability," you will quickly find that it's a common theme among paleoclimate researchers. Not only is the Milankovitch theory leading to the conclusion of an unstable climate, it is also leading to the conclusion that CO_2 changes were a major player in past climate change.

> Simulations with global climate models show that the amplitude [size] of glacial-interglacial temperature changes can only be reproduced if CO_2 changes are accounted for....This leads us to conclude that CO_2 changes are an important (feedback) factor in determining glacial-interglacial temperature changes although the ultimate cause of the ice age cycles are Earth's orbital cycles.[2]

Other paleoclimatologists agree:

> Amplification of orbitally forced insolation [sunlight] changes by greenhouse gases is a leading explanation for the large amplitude of late Pleistocene glacial-interglacial cycles.[3]

Notice the logic that secular scientists are using. They know that the changes in sunlight are too small in and of themselves to cause Ice Ages. But since they believe the Milankovitch theory is correct, they have concluded that carbon dioxide (and other factors) must somehow be amplifying these small changes in order to have brought about past Ice Ages. Moreover, they are also concluding that negative feedbacks, which could potentially mitigate the size of the supposed climate

"If the basis for the Astronomical Theory is wrong, many of the more radical environmental efforts may be unjustified."

change coming from changes in CO_2, must also be too small to effectively influence the climate. Hence, they believe climate sensitivity must be high (Figure 19.1). But this conclusion isn't coming from meteorology or even climatology per se. It is coming from the assumption that the Milankovitch theory is correct—in spite of the lack of evidence for the theory.

Secular Scientist "Know" the Astronomical Theory is Correct
(in spite of a lack of convincing evidence)

They *Also* Know Changes in Sunlight
are Too Small to be the Sole Cause of Ice Ages

Computer Models Can't Replicate Assumed Past Temperature Changes
Unless CO_2 is a Major "Positive Feedback"

CO_2 and Other Factors Are Greatly Amplifying Small Changes in Sunlight
Climate Sensitivity is High

Figure 19.1. The reasoning used by secular paleoclimatologists to conclude that climate sensitivity must be high

John Cook is a Ph.D. psychologist with a B.S. in physics. He is an assistant research professor at George Mason University's Center for Climate Change Communication and also runs the Skeptical Science blog, devoted to debunking arguments against global warming. Dana Nuccitelli is an environmental scientist and one of the website's contributors. In an article for the website, Nuccitelli tacitly affirms that uniformitarianism is contributing to climate alarmism:

> Thus when arguing for low climate sensitivity, it becomes difficult to explain past climate changes. For example, between glacial and interglacial periods, the planet's average temperature changes on the order of 6°C (more like 8-10°C in the Antarctic). If the climate sensitivity is low, for example due to increasing low-lying cloud cover reflecting more sunlight as a response to global warming, then how can these large past temperature changes be explained?[4]

How indeed? But what if the Milankovitch theory is wrong? And what if paleoclimatologists are incorrectly interpreting the chemical data in the ice cores (see chapter 17)? Nuccitelli goes on:

> The main limit on the [climate] sensitivity value is that it has to be consistent with paleoclimatic data. A sensitivity which is too low will be inconsistent with past climate changes—basically *if there is some large negative feedback which makes the sensitivity too low, it would have prevented the planet from transitioning from ice ages to interglacial periods,*

> for example. Similarly a high climate sensitivity would have caused more and larger past climate changes.[5]

In other words, Nuccitelli thinks climate sensitivity must be high because he believes oxygen isotope wiggles in ice and deep-sea cores represent dozens of previous Ice Age cycles. Creationists of course, disagree with that interpretation. But even within a uniformitarian framework, there are potential factors (discussed in the previous chapter) that can complicate the interpretation of oxygen isotope wiggles in both the seafloor sediments and the ice cores. Furthermore, the ages assigned to these wiggles is coming from the Milankovitch Ice Age theory. But as we have seen in chapter 9, there are enormous problems with this theory, and the evidence for it is weak at best and quite possibly nonexistent. Appendices C and D provide even more detailed discussions of this topic. So, creation scientists have concerns about the numbers depicted on both the vertical and horizontal axes in Figure 19.2. The timescales secular scientists have assigned to the cores are grossly in error, but the temperature change estimates could also be in serious error, too.

It is also worth noting that Nuccitelli argues that the uniformitarian interpretation of past climate data is the *main* constraint on climate sensitivity. He apparently thinks

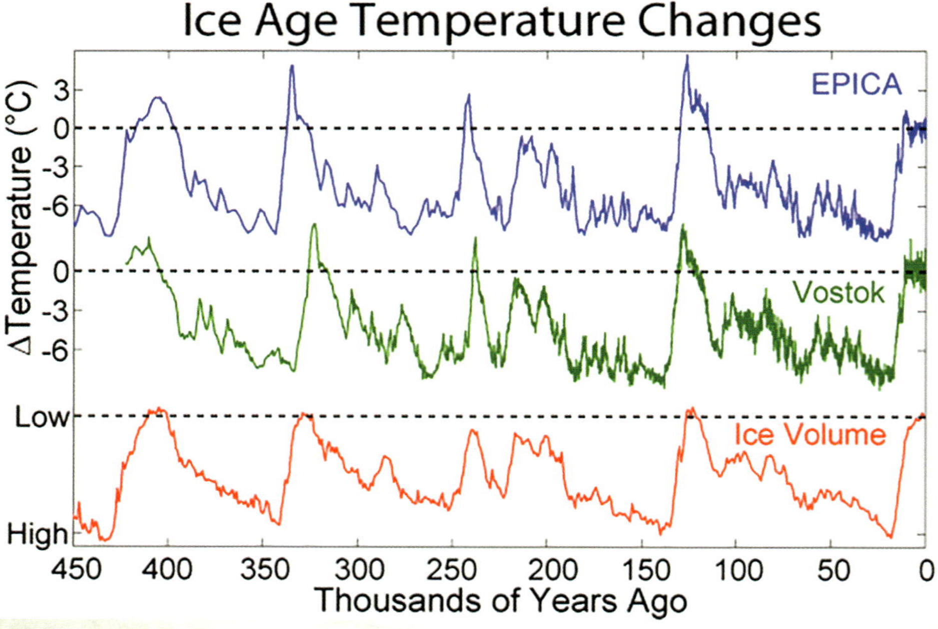

Figure 19.2. Inferred climate variables in sediment and ice cores are thought to tell a "story" of climate change over hundreds of thousands of years. At first glance, the apparent agreement and synchronization of these wiggle patterns seem to be a very strong argument for both an old earth and the secular Ice Age theory. However, appendix B explains why this is not the case.

“Global warming skeptic and former meteorologist Brian Sussman claims that this skepticism isn’t just limited to TV weathermen but is nearly universal among well-credentialed hurricane scientists and forecasters.”

it should be given the most weight when attempting to discern what that sensitivity is.

Of course, creation scientists would argue that a uniformitarian interpretation of these data should be given no consideration at all since that interpretation is fundamentally flawed.

By the way, it might not be a coincidence that meteorologists tend to be much more skeptical of global warming than climatologists.[6] Alarmists would argue that this skepticism is due to a lack of deep scientific knowledge (most meteorologists are not research scientists) and a focus on short-term rather than long-term trends. However, global warming skeptic and former meteorologist Brian Sussman claims that this skepticism isn’t just limited to TV weathermen but is nearly universal among well-credentialed hurricane scientists and forecasters.[7]

Could there be another reason for this disagreement? Could it be that paleoclimatologists are being heavily influenced by an incorrect interpretation of paleoclimate

data and they are allowing that dubious interpretation to trump sound meteorological and physical reasoning? In other words, many atmospheric scientists and meteorologists are focusing on the basics of atmospheric physics, which leads them to conclude that climate sensitivity is probably low. But the paleoclimatologists are focusing on a flawed interpretation of paleoclimate data, which is leading them to conclude that climate sensitivity is high.

Nuccitelli also uses results from a 2009 paper[8] to make an argument for a high climate sensitivity:

> One recent study examining the Palaeocene-Eocene Thermal Maximum (about 55 million years ago), during which the planet warmed 5-9°C, found that "At accepted values for the climate sensitivity to a doubling of the atmospheric CO_2 concentration, this rise in CO_2 can explain only between 1 and 3.5°C of the warming inferred from proxy records."...This suggests that climate sensitivity may be higher than we currently believe, but it likely isn't lower.[9]

There is strong evidence that nearly all the fossiliferous sedimentary rocks are the result of the Genesis Flood, including most Cenozoic rocks, with uniformitarian ages much less than 66 million years (the alleged time dinosaurs went extinct).[10,11] This means that these ostensibly 55-million-year-old sediments were deposited during the Flood itself. These were anything but "normal" conditions. So it is *totally* invalid to use these chemical wiggles to infer present-day climate sensitivity.

Inferring Past Sea Levels from Seafloor Sediments

According to NOAA, global sea level has risen about nine inches since 1880.[12] This translates into about 1.6 millimeters per year, which hardly seems to be a cause for concern. However, this rate has nearly doubled since 1993 to about 3 millimeters per year. There is much uncertainty in these estimates, and separating local from global effects is quite difficult, as climatologist Judith Curry explains:

> There is no question that local sea levels are increasing in some locations at rates that are causing damage in coastal regions. However attributing these changes to human caused global warming (even if you assume that recent global sea level rise is caused entirely by greenhouse gas emissions) is very challenging. This is because there are much larger impacts on local sea level rise from local geological processes, land use practices and coastal engineering.[13]

Furthermore, it is not a given that this trend, even if real, will continue.[14] So, is some other factor contributing to concern over rising sea levels? Yes, and as you may have already guessed, it's the Milankovitch theory. Oceanographer Wolfgang Berger wrote:

> Just when can we expect to see a rapid rise of sea level, ten times higher than the present values of a few millimeters per year? We do not know. All we can say, *from experience with the many millennia of the ice-age records in the deep sea*, is that once melting starts, it stimulates further melting for centuries. Deglaciation keeps going once begun in earnest: a great example of the dilemma of the sorcerer's apprentice.[15]

"There are much larger impacts on local sea level rise from local geological processes, land use practices and coastal engineering."

Note that Berger's concern about rapidly rising sea levels is based upon "experience with the many millennia of the ice-age records in the deep sea." What Berger is calling "experience" is really a Milankovitch interpretation of seafloor sediment oxygen isotope ($\delta^{18}O$) values. How could these oxygen isotope values lead secular scientists to conclude that future catastrophic flooding is a possibility?

Remember that secular scientists believe the sediment oxygen isotope values tell them that much ice was on the continents at different times in the "prehistoric" past. However, the mass of water on the earth's surface and in the earth's atmosphere is basically constant.[16] Also, the mass of water frozen as sea ice is much less than the amount of ice on the continents. This means that when more water is "locked up" as continental ice, global sea level drops. Likewise, when less water is "locked up" as ice, then global sea level rises. So, secular scientists see the oxygen isotope values as a measure of not just global ice volume but an indirect measure of global sea level.[17,18]

Also, remember that once uniformitarian scientists became convinced that the Milankovitch theory was correct, they began using it to assign ages to deep-sea sediment and ice cores. This was done via orbital tuning, in which the Milankovitch theory is assumed to be true and is used to assign ages to the oxygen isotope wiggles.

Since secular scientists think the oxygen isotope values give them a record of global sea level over time, these assigned ages allow them to calculate rates of sea level rise. For example, if you want to calculate the rate (or velocity) at which your car is traveling, you need both your change in position and the time it took for that change to occur. Dividing the change in position by the time gives the rate.

In the same way, secular scientists think that they can use the seafloor sediment data to calculate past rates of sea level rise. They think the oxygen isotope values give them estimates of past global sea levels, and the ages that have been assigned to the oxygen isotope wiggles allow them to calculate how quickly these changes in sea level occurred. Based on their calculations, some of these supposed sea level changes were quite fast.

But what if these calculated rates are wrong? These orbitally tuned ages are coming from the Milankovitch Ice Age theory. At the risk of sounding like a broken record, recall that the evidence for this theory is weak at best, even by secular reckoning. This means that these calculated rates of past sea level rise are suspect, even for someone who believes in an old earth. Hence, concern about rapidly rising sea levels can't really be justified from the seafloor sediment data.

Inferring Past Sea Levels from Fossil Corals

However, uniformitarian scientists also use other methods to infer past sea levels. For instance, some types of corals grow in shallow water but not in deep water.[19] For this reason, secular scientists think they can use fossil coral reefs to infer past changes in sea level. They also think they can use radioisotope dating (radiocarbon for "younger" fossil corals and uranium-thorium dating for "older" ones) to assign ages to these fossil corals. Sometimes they infer coral ages by comparisons with other ancient shorelines to which they have already assigned ages.[20] Of course, if past sea levels are known, as well as the dates for these past sea levels, then past rates of sea level rise can be calculated.

> "But there's an even bigger problem. Many of these fossil corals are located in Flood deposits."

However, there are two big problems with using fossil corals to infer past sea levels and rates of sea level rise. First, even uniformitarian scientists don't entirely trust uranium-lead dating.[21] Nor do they really trust radiocarbon dating, since published dates are actually *selected* dates.[22] In fact, the Pacemaker paper itself (discussed in chapter 9 and appendix C) is a good example of the arbitrary manner in which secular scientists use or don't use the radiocarbon method, depending on whether or not it is expedient for them to do so.[23]

But there's an even bigger problem. Many of these fossil corals are located in Flood deposits. Given the geologic upheaveal of the Flood, it's quite possible that many of these fossil corals are not even located in their original growth positions. In fact, fossil coral reefs are generally much smaller and less organized than living coral reefs. Creationists think that these smaller pieces of coral came from coral reefs that were broken up during the Flood and mixed with limestone deposits.[24] Obviously, this would invalidate any past sea levels inferred from these fossil reefs.

As an aside, uniformitarians have claimed that coral reefs invalidate the biblical timescale.[25] Supposedly, coral growth is so slow that the 1,656 years between creation and the Flood would not have been sufficient time for the growth of large corals in the pre-Flood world. Likewise, some scientists claim that some modern-day reefs are so large that they could not have attained their present sizes in the 4,500 years since the Flood. However, reef growth rates can

Fossil coral

"Large coral reefs such as Australia's Great Barrier Reef can plausibly grow within the biblical timescale."

vary greatly due to a number of factors.[26] Although creation researchers need to do more work in this area, they have already shown that large coral reefs such as Australia's Great Barrier Reef can plausibly grow within the biblical timescale.[27] They have also plausibly addressed what uniformitarians would likely consider their very strongest argument—that coral reefs prove an old earth.[28]

Secular researchers have concluded, based on their analysis of coral data, that sea level rise can be fast and erratic, even over a scale of just decades. For instance, a team of researchers from Rice University and Texas A&M-Corpus Christi reached that conclusion by analyzing fossil reefs in relatively shallow water in the Gulf of Mexico.[29] A press release from Rice University stated:

> Scientists from Rice University and Texas A&M University-Corpus Christi's Harte Research Institute for the Gulf of Mexico Studies have discovered that Earth's sea level did not rise steadily but rather in sharp, punctuated bursts when the planet's glaciers melted during the period of global warming at the close of the last ice age. The researchers found evidence in drowned reefs offshore Texas that showed sea level rose in several bursts ranging in length from a few decades to one century.[30]

The press release went on to say:

> Given that more than half a billion people live within a few meters of modern sea level, he [Rice geologist and study co-author Jeffrey Nittrouer] said punctuated sea-level rise poses a particular risk to those communities that are not prepared for future inundation.[31]

If rapid, erratic sea level rise can occur in just 10 years, then that obviously is cause for concern, especially for people living in low-lying coastal areas. But is it valid to compare today's climate with the end of the Ice Age?

Here is an example of how paleoclimate researchers are drawing false analogies between our present climate and the end of the Ice Age. Uniformitarians concluded that meltwater from rapid melting at the end of the last Ice Age 11,500 years ago

drowned these corals. In this particular case, that interpretation may very well be essentially correct, although creationists would argue that the secular age assignment is too high. An enormous amount of flooding *did* occur at the end of the post-Flood Ice Age (see chapter 13), and it is not hard to imagine that some of this meltwater could have drowned corals living in shallow water. But the Ice Age, including its end, was a unique time in Earth history. For this reason, it is invalid to use the end of the Ice Age to draw inferences about our present-day climate.

False Analogies

This is not the only example. Remember that secular scientists think that there have been 50 Ice Ages during the last 2.6 million years, with warmer interglacials between these Ice Ages. In their thinking, the current warm period is just one of dozens of warm periods that have occurred in the last 2.6 million years. In 1993, secular climate scientists concluded, based on their interpretation of chemical data from the Greenland ice cores, that climate change could occur very rapidly:

> [The] climate in Greenland during the last interglacial period was characterized by a series of severe cold periods, which began extremely rapidly and lasted from decades to centuries. As the last interglacial seems to have been slightly warmer than the present one, its unstable climate raises questions about the effects of future global warming.[32]

Note their logic. They claimed that the previous interglacial was characterized by severe climate change. Since we are supposedly in another interglacial, they argue by analogy that our present interglacial could *also* experience severe climate change.

But in the creation model, these $\delta^{18}O$ wiggles within the Greenland ice cores were all deposited within the last 4,500 years. There was no previous interglacial, only the post-Flood Ice Age and the post-Ice Age climate that followed it. Hence, this argument by analogy is invalid.

Some creation scientists think that severe cold periods may have indeed occurred at the end of the Ice Age.[33] But as we have already shown, the Ice Age was caused by the Genesis Flood, a never-to-be-repeated event. Because it was a unique time in Earth history, it is logically invalid to use this unique "transitional" climate to draw conclusions about future climate change.

To be fair, secular scientists may have since changed some of their conclusions from this 1993 paper.[34] But I mention this example to illustrate how secular scientists use an old-earth interpretation of climate data to draw conclusions about future climate change. This last, particular example may no longer be valid, but uniformitarian scientists still use this kind of reasoning to draw conclusions about future climate change. Obviously, if their uniformitarian philosophy is in error, their conclusions will be too! What you believe about the past really does affect what you believe about the future.

The rainbow is a reminder that the Flood is a never-to-be-repeated event (Genesis 9:11-17)

References

1. Vardiman, L. 2001. *Climates Before and After the Genesis Flood: Numerical Models and Their Implications* El Cajon, CA: Institute for Creation Research, 79.
2. Schmittner, A. Introduction to Climate Science. Online text for Oregon State University, College of Earth, Ocean, and Atmospheric Sciences. Posted on library.open.oregonstate.edu, accessed January 4, 2019.
3. Cronin, T. M. 2010. *Paleoclimates: Understanding Climate Change Past and Present.* New York: Columbia University Press, 125.
4. Nuccitelli, D. How sensitive is our climate? Skeptical Science. Posted on skepticalscience.com, accessed November 13, 2018.
5. Ibid, emphasis added.
6. Kaufman, L. Among Weathercasters, Doubt on Warming. *New York Times.* Posted on nytimes.com March 29, 2010 accessed December 7, 2018.
7. Sussman, B. 2010. *Climategate.* Washington, DC: WND Books, 138-139.
8. Zeebe, R. E., J. C. Zachos, and G. R. Dickens. 2009. Carbon dioxide forcing alone insufficient to explain Palaeocene-Eocene Thermal Maximum warming. *Nature Geoscience.* 2: 576-580.
9. Nuccitelli, How sensitive is our climate?
10. Holt, R. D. 1996. Evidence for a late Cainozoic Flood/post-Flood boundary. *Journal of Creation.* 10 (1): 128-167.
11. Oard, M. J. 2007. Defining the Flood/post-Flood boundary in sedimentary rocks. *Journal of Creation.* 21 (1): 98-110.
12. Lindsey, R. Climate Change: Global Sea Level. Posted on climate.gov.
13. Curry, J. Sea level rise acceleration (or not): Part VII. U.S. Coastal Impacts. Posted on judithcurry.com April 15, 2018, accessed November 13, 2018.
14. Spencer, R. W. Sea Level Rise: Human Portion is Small. Posted on drroyspencer.com May 25, 2018, accessed November 13, 2018.
15. Berger, W. H. 2012. Milankovitch Theory—Hits and Misses. Scripps Institution of Oceanography, University of California San Diego, La Jolla, California, 16. Emphasis added.
16. Volcanoes, for instance, put water vapor into the atmosphere when they erupt. However, this is a small amount compared to the total amount of water on the earth's surface and in the atmosphere.
17. Waelbroeck, C. et al. 2002. Sea-level and deep water temperature changes derived from benthic foraminiferal isotopic records. *Quaternary Science Reviews.* 21: 295-305.
18. Raymo, M. E. et al. 2018. The accuracy of mid-Pliocene $\delta^{18}O$-based ice volume and sea level reconstructions. *Earth-Science Reviews.* 177: 291-302.
19. Woodroffe, C. and J. M. Webster. 2014. Coral Reefs and Sea-Level Change. *Marine Geology.* 352: 248-267.
20. Khanna, P. et al. 2017. Coralgal reef morphology records punctuated sea-level rise during the last deglaciation. *Nature Communications.* 8: 1046.
21. Andersen, M. B. et al. 2009. U-series dating of fossil coral reefs: Consensus and controversy. M. Siddall et al, eds. *PAGES News.* 17 (2).
22. Morris, J. D. 1994. *The Young Earth.* Colorado Springs, CO: Master Books, 65.
23. Hebert, J. 2016. Revisiting an Iconic Argument for Milankovitch Climate Forcing: Should the "Pacemaker of the Ice Ages" Paper Be Retracted?—Part 1. *Answers Research Journal.* 9: 31.
24. Morris, J. D. 2013. Fossil Coral 'Reefs' Among Rock Strata. *Acts & Facts.* 42 (12): 13.
25. Weber, C. G. 1980. The Fatal Flaws of Flood Geology. *Creation/Evolution Journal.* 1 (1): 24-37.
26. Thomas, B. 50-Year Study Shows Coral 'Clocks' Unreliable. *Creation Science Update.* Posted on ICR.org January 28, 2011, accessed December 18, 2018.
27. Read, P. and A. Snelling. 1985. How old is Australia's Great Barrier Reef? *Creation.* 8 (1): 6-9.
28. Oard, M. 1999. The paradox of Pacific guyots and a possible solution for the thick 'reefal' limestone on Enewetok Island. *Journal of Creation.* 13 (1): 1-2.
29. Khanna, Coralgal reef morphology records punctuated sea-level rise during the last deglaciation.
30. Boyd, J. and D. Ruth. Fossil coral reefs show sea level rose in bursts during last warming. Rice University press release. Posted on news.rice.edu October 19, 2017, accessed December 18, 2018.
31. Ibid.
32. Greenland Ice-Core Project (GRIP) Members. 1993. Climate Instability During the Last Interglacial Period Recorded in the GRIP Ice Core. *Nature.* 364: 203-207.
33. Oard, M. J. 2005. *The Frozen Record.* El Cajon, CA: Institute for Creation Research, 125-128.
34. GRIP members, Climate Instability During the Last Interglacial Period Recorded in the GRIP Ice Core.

20 Conclusion

Summary: Since the Bible is telling the truth about the early history of our planet, we can't leave it out of the climate change debate. Doing so leads to incorrect conclusions about past climate change, as well as the earth's age. We have good scientific and biblical reasons to think that Earth's climate is stable. The human race is not facing a climate emergency. The real emergency is the coming judgment of God on sin. Every person will have to give an account to God for the wrong he or she has done. Just as Noah's Ark was the only way to escape the global Flood, the Lord Jesus Christ is the only way to escape the coming judgment.

In this book I have attempted to methodically present the case that the Bible provides a much better framework for explaining past climate change than the uniformitarian story. Likewise, I have attempted to show that a uniformitarian worldview is driving much of climate change, or global warming, alarmism.

You may have noticed that I haven't really addressed whether or not I personally think global warming is occurring. There is a reason for this. Even if global warming *is* occurring, it's not something we should panic about. Draconian climate regulations are not needed to forestall a climate catastrophe. Remember, the key issue in this debate is the stability of the earth's climate system. If our climate system is stable, then it will self-adjust to prevent out-of-control warming. I would assert that the arguments for an unstable climate are invalid. They are coming from dubious parameters in computer models and uniformitarian interpretations of paleoclimate data. In some cases, the arguments proceed from mishandling (and perhaps even fraudulent handling) of the data. On the other hand, we have good theological and scientific reasons to expect Earth's climate to be relatively stable.

For the record, I strongly suspect that global warming is *not* occurring now. This is partly because of the well-known warming pause that has been occurring for the last

20 years or so. But even if warming is occurring right now, it will not continue indefinitely, and it is not a reason to panic.

What conclusions can we draw from this study? For non-Christians, the reality of the Genesis Flood should be a sober warning that God does indeed judge sin. The Bible makes it very clear that God will judge the world again, not with water, but with fire.

> ...when the Lord Jesus is revealed from heaven with His mighty angels, in flaming fire taking vengeance on those who do not know God, and on those who do not obey the gospel of our Lord Jesus Christ. These shall be punished with everlasting destruction from the presence of the Lord and from the glory of His power, when He comes, in that Day, to be glorified in His saints and to be admired among all those who believe. (2 Thessalonians 1:7-10)
>
> Then I saw a great white throne and Him who sat on it, from whose face the earth and the heaven fled away. And there was found no place for them. And I saw the dead, small and great, standing before God, and books were opened. And another book was opened, which is the Book of Life. And the dead were judged according to their works, by the things which were written in the books. The sea gave up the dead who were in it, and Death and Hades delivered up the dead who were in them. And they were judged, each one according to his works. Then Death and Hades were cast into the lake of fire. This is the second death. And anyone not found written in the Book of Life was cast into the lake of fire. (Revelation 20:11-15)

"If our climate system is stable, then it will self-adjust to prevent out-of-control warming. I would assert that the arguments for an unstable climate are invalid."

We live in a society that takes offense at the very idea that God has the right to hold us accountable for our actions. When calamity occurs, the skeptics mockingly ask, "Where was God?" or "How could God allow this to happen?" However, all the evil in our world is ultimately humans' fault, not God's. The bad things that occur are a consequence of mankind's sin. This includes murder and violence, but it also includes natural evil such as earthquakes and volcanoes. These too are ultimately the result of our sin, the aftereffects of the Flood of Noah. God is under no obligation whatsoever to spare us from the consequences of our actions.

Furthermore, this judgment is not limited to this world. God has made provision for our sins in the Person of Jesus Christ. He has made it possible for our sins to be forgiven and for us to be adopted into His family. Christ, the Son of God, the Second Person of the Godhead, became a man and lived a perfect life. Because He never sinned, He was able to pay the penalty for our sins on the cross. Then He was buried, and God raised Him from the dead.

The Lord Jesus Christ paid the penalty for our sins on the cross

The Bible promises that if we believe in Christ and turn from our sins, we will be saved.

> For God so loved the world that He gave His only begotten Son, that whoever believes in Him should not perish but have everlasting life. (John 3:16)
>
> But what does it say? "The word is near you, in your mouth and in your heart" (that is, the word of faith which we preach): that if you confess with your mouth the Lord Jesus and believe in your heart that God has raised Him from the dead, you will be saved. For with the heart one believes unto righteousness, and with the mouth confession is made unto salvation. (Romans 10:8-10)

However, if we reject Christ, there is no other remedy.

> He who believes in the Son has everlasting life; and he who does not believe the Son shall not see life, but the wrath of God abides on him. (John 3:36)

Just as Noah's Ark was the only way to escape the judgment of the great Flood, the Lord Jesus is the only way of salvation.

> Nor is there salvation in any other, for there is no other name under heaven given among men by which we must be saved. (Acts 4:12)

If you are not a believer in Christ, I hope this book challenges you. It is my hope and prayer that you are convinced of the reality of the Genesis Flood and the coming future judgment. It is my prayer that you will realize that you are *not* "basically a good person." Instead, you are a sinner in need of God's grace. It is also my prayer that you will recognize that you are confronted with a more serious emergency than climate change—namely, your need for salvation.

For Christians, I hope this book has made you think. For those evangelicals who are frustrated by the refusal of fellow evangelicals to jump on the climate change bandwagon, I hope this book has given you some insight into the reasons for our skepticism. I hope it has also caused you to question whether the issue of climate change is as urgent as some would have you believe. In light of the infinitely more pressing need to proclaim the gospel of Jesus Christ to a dying world, is it wise for Christians

> "If the Bible is telling the truth about the early history of our planet, then we cannot leave the Bible out of the climate change debate. Furthermore, we cannot ignore what the Bible says about the earth's age."

to allow themselves to be distracted from the Great Commission by secondary issues such as climate change?

I also hope this book will cause evangelicals to think twice before accepting secular stories about the past, stories that clearly contradict God's Word. One cannot hope to correctly interpret past climate data using an incorrect paradigm of Earth history. If the Bible is telling the truth about the early history of our planet, then we cannot

leave the Bible out of the climate change debate. Furthermore, we cannot ignore what the Bible says about the earth's age. Many Christians long ago realized that the theory of evolution was intellectually bankrupt, yet they are still, for some reason, hesitant to challenge the doctrine of an old earth. While dodging tough questions about Earth's age and history might be easier in the short run for the Christian apologist, it is a serious mistake. I doubt that many non-Christians are impressed by such an avoidance tactic. Unbelievers have no difficulty seeing the inherent inconsistency exhibited by Christians who fail to take seriously the first 11 chapters of their own holy book.

Furthermore, there is no good reason to dodge these issues. Creation scientists have made tremendous strides in answering objections to a literal understanding of Genesis. They have already done the heavy lifting in showing that biblical creation is a viable alternative to the secular model. In fact, I have no hesitation saying that the creation model is vastly superior to the evolutionary model. And this is true not just with respect to the origin of life, but with respect to geology, climate change, and the age of the earth!

Is There a Connection Between Weather and Climate, Cosmic Rays, and Solar Activity?

Summary: Climate computer models are a major source of climate change alarmism because they predict extreme warming due to increased atmospheric carbon dioxide. However, many aspects of the climate are still poorly understood, and these models necessarily fail to take those factors into account. Clouds are a major source of uncertainty in the global warming debate. Scientists have long speculated on a possible link between solar activity, cosmic rays, and weather and climate, and the connection between these factors may have already been found. However, the true explanation is likely more complicated than some believe and involves more than just cosmic rays and solar activity. If this connection is confirmed, then incorporating these effects into climate models could help improve both short-term weather forecasts and estimates of climate sensitivity.

Scientists have long speculated that the sun could somehow be affecting Earth's weather and climate. As far back as 1801, famed astronomer William Herschel published a paper that suggested the number of sunspots was somehow influencing wheat prices.[1]

Sunspots are small regions on the sun's surface that are much cooler than the surrounding surface. The number of sunspots varies with the sun's level of activity. The sun is less active when it has few sunspots, and there is a slight decrease in the sun's energy output.

Most people think it's obvious that the sun would affect weather and climate. After all, Earth is continually bathed in sunlight. However, such a connection is not as obvious as one might think. In order for the sun to cause changes in Earth's weather and climate, something about the sun has to change. Yet, the energy that the earth receives from the sun changes very little over time. For this reason, many scientists have dismissed the suggestion that the sun could somehow be influencing weather or climate here on Earth.

However, the sun's magnetic field does have an influence on the number of *galactic cosmic rays* (GCRs) entering our atmosphere. The name is somewhat misleading, because galactic cosmic rays aren't rays at all but instead are energetic particles (mainly positively charged protons) originating from within our own Milky Way galaxy. Magnetic fields can cause moving charged particles to turn, and the sun's magnetic field, which extends outward into the rest of the solar system, has an influence on how many of these charged protons enter our atmosphere. During times of low solar activity (few sunspots), more cosmic rays enter the earth's atmosphere, but during times of high solar activity (more sunspots), fewer cosmic rays enter our atmosphere. In fact, during an 11-year sunspot cycle (Figure A.1), during which the number of sunspots goes from one minimum value to another minimum, the number of GCRs entering our upper atmosphere can vary by as much as 15%.[2] By contrast, the amount of energy from the sun varies by less than 0.1% over the 11-year solar cycle.[3]

Scientists have long speculated that the sun could somehow be influencing weather and climate here on Earth. The number of sunspots on the sun is a proxy or "stand in" for the sun's activity. Fewer sunspots are present when the sun is less active, and more sunspots are present when it is more active.

When GCRs collide with atoms in our atmosphere, showers of subatomic particles are produced, ultimately resulting in the production of a large number of *ions*

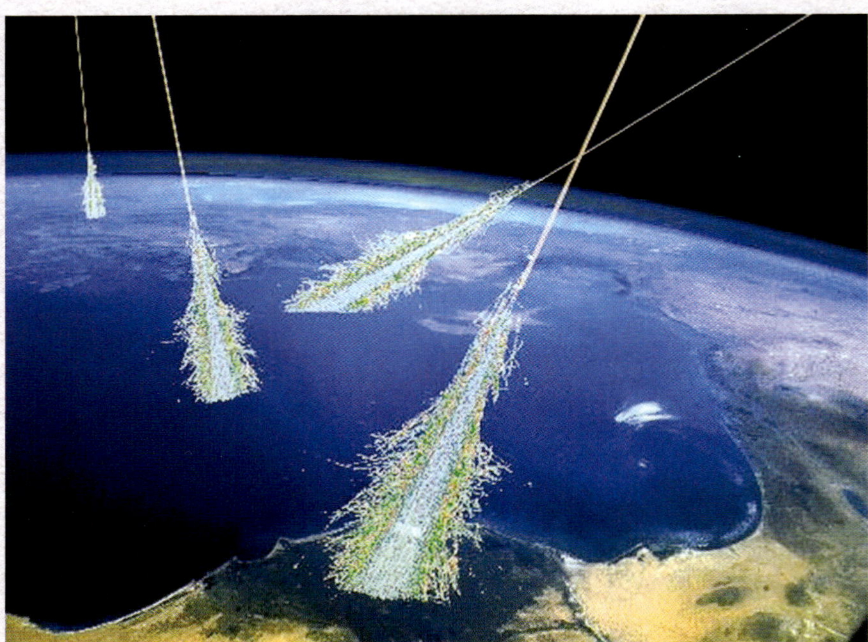

When cosmic rays (fast-moving protons) enter the atmosphere, charged particles called ions are produced. The sun has a strong influence on the number of cosmic rays entering the atmosphere and therefore on the number of atmospheric ions, which could conceivably affect cloud properties. This in turn could affect weather and climate. More cosmic rays enter the atmosphere when solar activity is low.

(charged particles) within our atmosphere. So, galactic cosmic rays would seem to be a natural candidate for the mechanism by which the sun is influencing weather on Earth. In fact, ionization due to cosmic rays is apparently the *only* process in our lower atmosphere that is known to undergo large variations due to solar activity.[4]

In fact, the number of GCRs entering our atmosphere varies over shorter timescales as well. The sun sometimes ejects large bubbles of plasma (ionized gas) called *coronal mass ejections*. These plasma bubbles can decrease the number of GCRs entering the atmosphere over much shorter intervals (hours to days). These short-term decreases in GCRs entering our atmosphere are called *Forbush decreases*.

Ion-Mediated Nucleation

Currently, there are two main hypotheses about how cosmic rays could be influencing Earth's climate.[5] The first is an idea called *ion-mediated nucleation* (IMN) advocated by Danish physicist Henrik Svensmark. ICR has published past articles discussing Svensmark's idea.[6,7] Tiny particles and droplets called *aerosols* are present

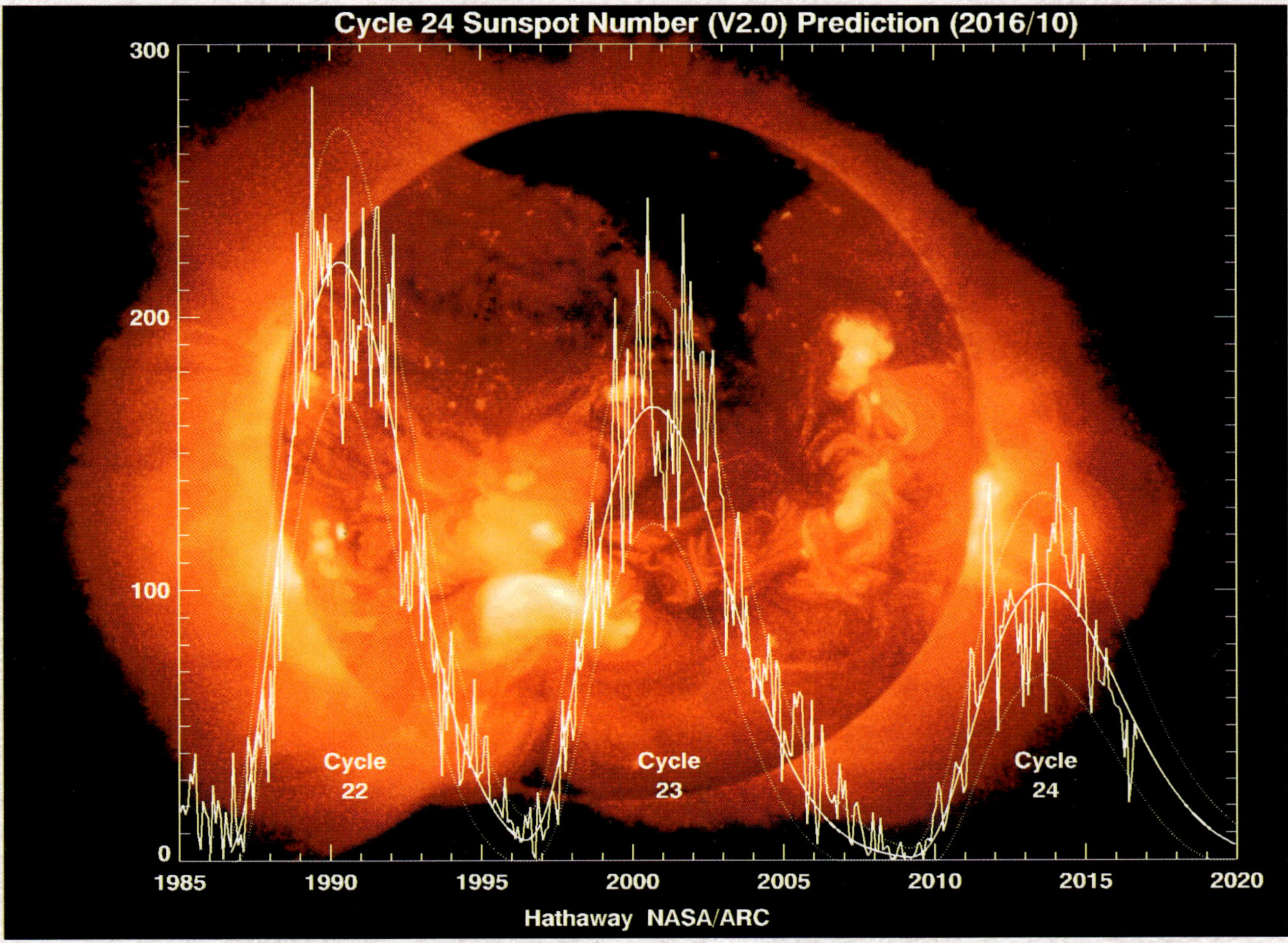

Figure A.1. The sun exhibits a regular 11-year sunspot cycle

in the atmosphere. Some of these aerosols have properties that make them suitable "seeds" for cloud droplets. These seeds are called *cloud condensation nuclei* (CCNs). In order for water vapor to form cloud droplets, the water vapor must condense onto the CCNs.[8]

The presence of charged particles (ions) in the atmosphere encourages the formation and growth of these cloud condensation nuclei. And since more cosmic rays produce more atmospheric ions, Svensmark argues there should be more clouds during times when more cosmic rays are entering the atmosphere. That's why his hypothesis is called ion-mediated nucleation: cosmic rays generate more ions in the atmosphere, which facilitate, or mediate, the growth of cloud condensation nuclei.

Although Svensmark's ideas have received a great deal of publicity on the internet, they have problems. One big problem is that Svensmark has had difficulty obtaining repeatable results. For instance, Svensmark and his colleagues have published multiple papers showing that there are more clouds when the number of cosmic rays increases. At first glance, this would seem to be an argument in favor of Svensmark's mechanism. However, in one study it was the total number of clouds over the oceans that tracked with the number of cosmic rays.[9] In other studies, it was the number of low-altitude clouds.[10,11] So, cosmic rays might *somehow* be affecting cloud cover. But are low-altitude clouds being affected or oceanic clouds? And since cosmic rays are more abundant in the upper atmosphere than in the lower atmosphere, one would expect high-altitude clouds to be more affected than low-altitude clouds. Yet, to the best of my knowledge, Svensmark has not published a paper showing a connection between high-level clouds and cosmic rays. In any case, he has not shown a consistent, repeatable pattern. The fact that Svensmark cannot obtain consistent predictable results strongly suggests that his mechanism is too simplistic to account for all the data.

Also, the IMN mechanism may simply not be strong enough to account for the observations. Advocates of the IMN mechanism have pointed to the results of an experiment performed at CERN that seemed to provide support for the idea.[12,13] However, even those who performed the experiment acknowledged that the results didn't make clear how many of the nucleated particles could grow large enough to serve as CCNs, nor did it simulate the complexities of our atmosphere.[14] Others have claimed that the IMN mechanism is about a 100 times too weak to account for observed changes in cloud properties, although Svensmark has disputed this.[15,16]

Charge Modulation of Aerosol Scavenging

That brings us to the second mechanism for how cosmic rays could affect Earth's climate, called *charge modulated aerosol scavenging* (CMAS), advocated by physicist Brian Tinsley. I personally think Tinsley's mechanism is much more convincing than Svensmark's, but I could be biased since Tinsley was my Ph.D. advisor at the University of Texas at Dallas.

When charged aerosols and water droplets collide, they almost always "stick together."[17] Tinsley's hypothesis involves the fact that cloud droplets can scavenge or absorb airborne aerosols, some of which can act as CCNs (Figure A.2). However, the rates at which cloud droplets can scavenge these aerosols are influenced by electrical charges that are present on both the cloud droplets and the aerosols. Sophisticated computer simulations have shown that the presence of like electrical charges on both droplets and aerosols increase the rates at which larger CCNs get absorbed by the droplets.[18] For those who remember from high school physics that "like charges repel," this might seem counterintuitive. Shouldn't the presence of like charges push the CCNs and droplets farther apart, decreasing the rates at which cloud droplets scavenge CCNs?

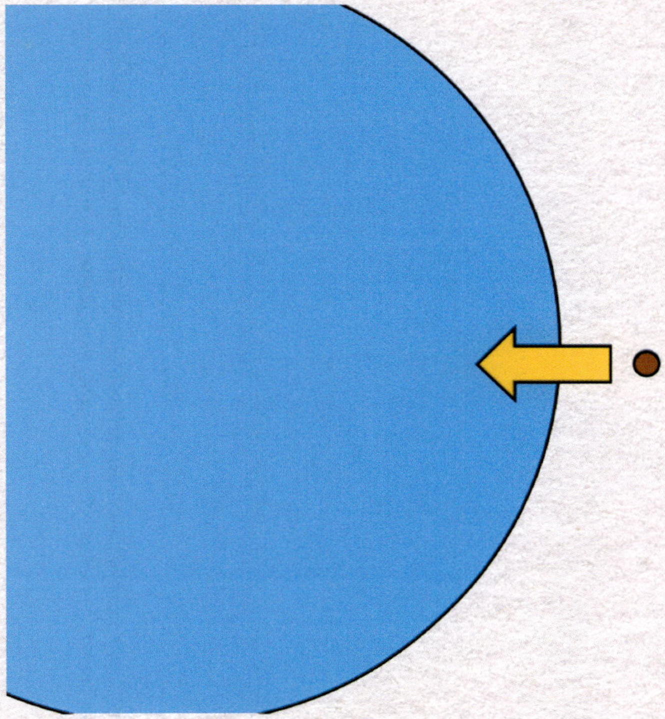

Actually, no. Cloud droplets and aerosols often contain conducting acids or salts.[19,20] For this reason, they are more like small conducting spheres than the simple "point charges" you learned about in high school. The charges distribute themselves on the droplets and CCNs in such a way that there are both repulsive *and* attractive electrostatic forces between them. The repulsive force is stronger when the spheres are farther apart, and the attractive forces are stronger when they are closer together.[21]

Figure A.2. Water droplets can scavenge or "gobble up" cloud condensation nuclei (CCNs), which are typically much, much smaller than cloud water droplets. The presence of electrical charges on the CCNs and droplets affects the rates at which this happens. Droplet and CCN are not drawn to scale.

Remember that computer simulations have shown that the presence of like charges on the CCNs and cloud droplets increase the rates at which the droplets scavenge or absorb the larger CCNs. They also decrease the rates at which the smaller CCNs get scavenged (Figure A.3). This means an increase in the percentage of freely floating small CCNs. This favors the formation of a greater number of smaller cloud droplets because larger CCNs, due to their larger sizes, have a "head start" on droplet growth over the smaller CCNs. Their smaller sizes mean that more water must condense onto a smaller CCN for it to attain just the starting size of a larger CCN. So, increasing the percentage of smaller CCNs favors the formation of smaller droplets.[22]

Rain occurs when collisions between large and small droplets produce droplets

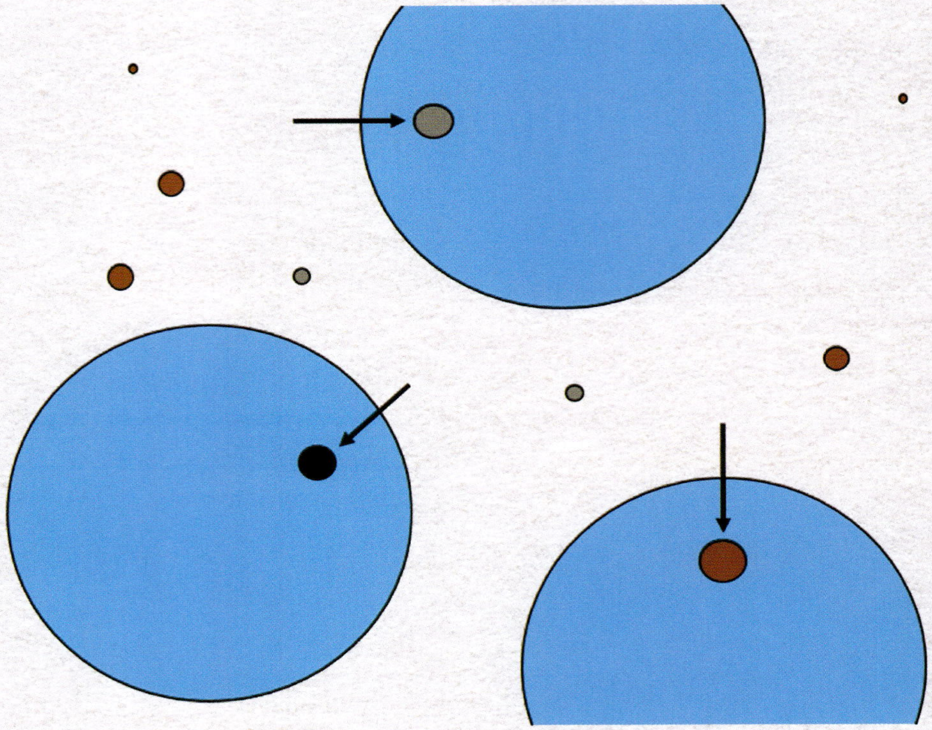

Figure A.3. Sophisticated computer simulations have shown that the presence of like charges on both CCNs and cloud droplets increases the rates at which the larger CCNs are scavenged by the cloud droplets. It also decreases the rates at which smaller CCNs are scavenged. This increases the percentage of small CCNs within the cloud, which favors the formation of more droplets of smaller sizes. This decreases the average droplet size within the cloud.

that are large enough to start falling. A wide range of droplet sizes makes this process more likely. Increasing the number of smaller droplets tends to narrow the range of droplet sizes, making rain less likely. But less rain means that clouds will persist for longer times. So, if some mechanism could somehow increase the number of like charges on droplets and CCNs, this would cause an increase in cloud cover. Tinsley thinks that mechanism is something called the *global electric circuit*.

A Review of High School Physics

Before explaining this mechanism, let's review some basic concepts about electricity. You may remember the equation V = IR from high school physics. This equation states that the electrical current I (measured in amperes) passing through an electrical conductor depends upon both the voltage V (measured in volts) driving the current and the electrical resistance R (measured in ohms) of the conductor. Voltage is somewhat like a difference in pressure. Just as an imbalance of pressure can drive a water current from one place to another, a voltage can drive an electrical current from one place to another.

If the electrical resistance stays the same, then a higher/lower voltage will cause the current I to increase/decrease. Likewise, if the voltage V stays the same, then an increase/decrease in R will cause the current I to decrease/increase. This relationship is necessary to understand how the CMAS mechanism works.

The Global Electric Circuit

Both the earth itself and the charged layer of the atmosphere called the *ionosphere* are very good electrical conductors. One can think of them as conducting plates of a capacitor. A capacitor is made of two conducting objects (usually metals) separated by a short distance. When a voltage, or potential difference, is applied across the conductors, one of the conductors stores positive electrical charge, and the other conductor stores negative electrical charge. The conductors don't necessarily have to be any particular shape, but high school physics classes often discuss parallel plate capacitors, which consist of two parallel metal plates separated by a small distance.

However, one can also construct a capacitor that is shaped like a sphere. If one puts a layer of insulation around a conducting metal ball and then places a metal shell on top of the insulation, the ball and shell can also act as plates of a spherical capacitor.

The ionosphere and surface of the earth can be thought of as plates of a spherical capacitor (Figure A.4). In a perfect capacitor, the charges on the ionosphere and ground would stay put. However, the atmosphere is not a perfect insulator. Because of charged particles (ions) in the atmosphere, the atmosphere weakly conducts electricity. Moreover, there is typically a voltage of about 250,000 volts between the ionosphere and the surface of the earth, although this voltage can be higher or lower at different times. This voltage drives a current downward from the ionosphere to the earth's surface.

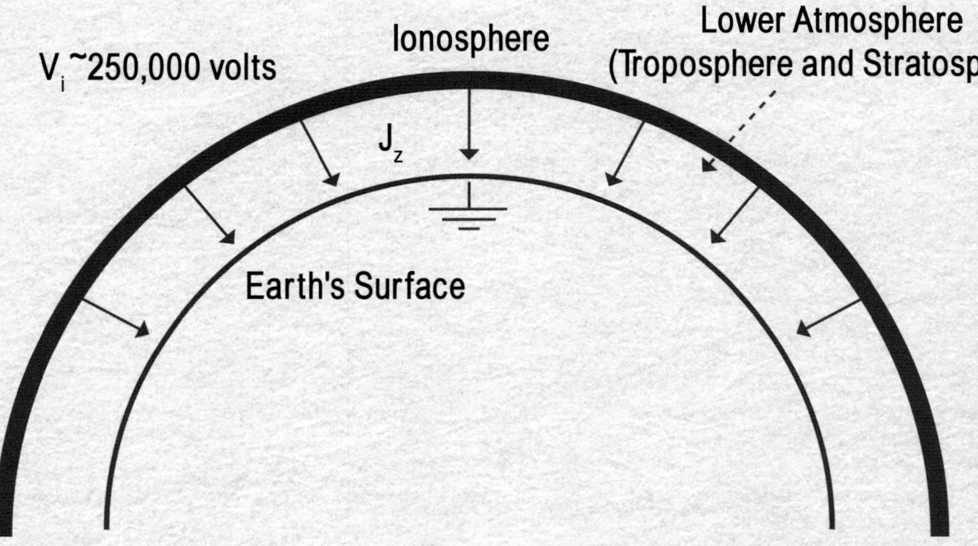

Figure A.4. The ionosphere and Earth's surface can be thought of as the conducting "plates" of a spherical capacitor. Because the atmosphere weakly conducts electricity, the voltage between the plates drives a tiny fair-weather current per unit area, or current density. Scientists designate this current density, measured in trillionths of an ampere per square meter, with the symbol J_z.

This current is not lightning. This is a fair-weather electrical current of about 1,000 amperes.[23] However, this current is spread out over the earth's surface so that the current falling on a square meter of the ground is tiny, about one to six trillionths of an ampere.[24] This current per square meter is called a *current density*, often denoted with the symbol J_z. The "J" is the typical letter physicists use to denote an electrical current density, and the "z" is often used to indicate the vertical direction. So, this current density is travelling downward in the vertical (or z) direction from the ionosphere to the earth's surface. This means that if you go outside on a clear day, there is actually a tiny current density coming down from the ionosphere that you could measure with sensitive instruments.

Clouds are much poorer conductors of electricity than the surrounding air.[25] When this fair-weather current density J_z passes through a cloud, the difference in electrical conductivity between the air and cloud causes a layer of positive charge to form on the cloud top and a layer of negative charge to form at the cloud bottom (Figure A.5). Tinsley believes these charges influence the rates at which cloud droplets scavenge CCNs. This in turn affects the number of clouds. As it turns out, the amount of this charge depends on the fair-weather current density J_z. A higher fair-weather current density J_z means more charge, which should result in more clouds. Likewise, a lower fair-weather current density J_z results in less charge, which should result in fewer clouds.

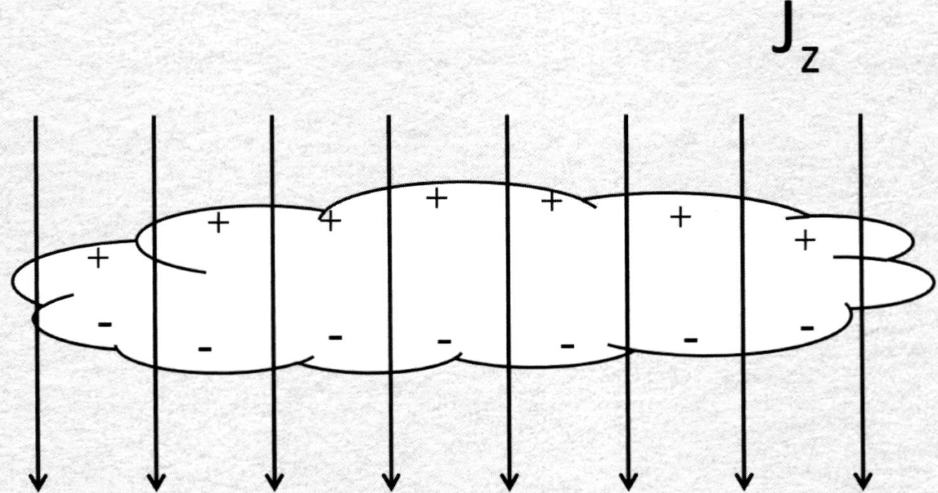

Figure A.5. When the current density J_z passes through a cloud, basic physics implies that a layer of positive charge forms on the cloud top and a layer of negative charge forms on the cloud bottom

Factors That Affect J_z

Just as the size of the current I depends upon R, the size of J_z is affected by the local electrical conductivity of the atmosphere. As it turns out, one of the factors influencing the conductivity of the atmosphere is the number of ions in the atmosphere, which in turn is affected by...galactic cosmic rays! A greater number of cosmic rays increases the number of atmospheric ions, which makes the atmosphere a better conductor of electricity, which increases J_z. Likewise, a smaller number of cosmic rays results in fewer atmospheric ions, which increases the electrical resistance of the atmosphere, which causes J_z to decrease.

> "As it turns out, one of the factors influencing the conductivity of the atmosphere is the number of ions in the atmosphere, which in turn is affected by... galactic cosmic rays!"

However, not just cosmic rays affect J_z. The size of the voltage between the ionosphere and the surface of the earth also affects J_z. As noted earlier, this voltage is typically 250,000 volts or so, but it can go up or down. Likewise, three other factors can influence J_z.[26]

The fact that J_z depends on factors besides just cosmic rays helps explain why Svensmark has had difficulty obtaining consistent, repeatable results. At times when the other factors affecting J_z stay constant, an increase in the number of cosmic rays will cause J_z to increase, which will put more charge on cloud droplets and CCNs, which will result in more cloud cover. This is consistent with Svensmark's observations. However, this tracking of cloud cover with cosmic rays will only hold so long as the other factors don't change. But these other factors *do* change over time, so Svensmark's results were dependent upon him fortuitously picking a time interval during which the other factors influencing J_z didn't change much. When those other factors changed, the apparent correlation between cosmic rays and cloud cover disappeared.

This means that the true quantity of interest is not the number of galactic cosmic rays entering the atmosphere per se, but rather the fair-weather current density

J_z. Cosmic rays are just one of multiple factors affecting J_z, which in turn is affecting cloud cover.

Evidence for a Solar-Cosmic Ray/Weather Connection

The evidence for Tinsley's working hypothesis is impressive. Space doesn't permit a detailed discussion, but I have published a series of four technical papers that elaborate on this evidence and address anticipated objections to the CMAS mechanism. These articles may freely be read online; active links are found at icr.org/jake_hebert.[27-30] Factors that cause J_z at a given location to increase have repeatedly been associated with greater cloud cover at that location. Likewise, factors that cause J_z to decrease have repeatedly been shown to be associated with less cloud cover (Figure A.6).

Figure A.6. Factors (including a smaller number of cosmic rays) that cause J_z at a particular location to become smaller have repeatedly been shown to "track" (or correlate) with less cloud cover at that location. Likewise, factors that cause J_z at some location to become larger have repeatedly been shown to track with greater cloud cover.

At the time I began working for him, Dr. Tinsley had already worked on this problem for about 20 years and had already concluded, via a chain of physical reasoning, that increases in J_z would increase winter cyclone intensity. He immediately tasked me with attempting to see if I could find evidence for such an effect. It was tremendously hard work extracting a faint signal from the data, but my Ph.D. research ultimately yielded evidence for this effect from two independent data sets.[31,32] I also found preliminary evidence for this effect from a third data set, although it was never

A low-pressure cyclone near Iceland. Although available data are limited, higher values of J_z have repeatedly been shown to be associated with more intense winter cyclones in the northern high latitudes.

formally published. I didn't have time to do a more rigorous analysis since I was hired by the Institute for Creation Research around that time. All three of these results were presented during a poster session at the 2011 meeting of the American Geophysical Union.[33]

A Connection to Very Cold European Winters?

Interestingly, this intensification of high-latitude winter cyclones may be the key to explaining particularly cold winters in Europe during times of very low solar (or low sunspot) activity. The so-called Maunder Minimum was a period (1645–1715 AD) when very few sunspots were observed on the surface of the sun. It coincided with the coldest part of the Little Ice Age. At that time, temperatures in Europe were so cold that normally ice-free rivers froze, including the Thames in London. In fact, the ice was thick enough that people had "frost fairs" on the ice![34]

Cyclones often form in the area between Iceland and Greenland. Because cyclones are low-pressure weather systems, this area in the North Atlantic is known as the Icelandic Low. Studies have repeatedly shown that when cyclones in the North Atlantic become more intense, there is an accompanying intensification of high-pressure weather systems to the east of ("downstream" from) the cyclones. The effect is greatest during the winter and during times of low solar activity.[35,36]

"The so-called Maunder Minimum was a period (1645–1715 AD) when very few sunspots were observed on the surface of the sun. It coincided with the coldest part of the Little Ice Age."

Such high-pressure systems can often remain nearly stationary and are called *blocks* or *blocking events*. They are often associated with severe weather if they persist for significant durations.[37] Interestingly, blocking events have been shown to last longer, be more intense, and be located farther east during periods of low solar activity. The effect may depend on the phase (east or west) of what meteorologists call the *Quasi-Biennial Oscillation* (QBO).[38]

The CMAS mechanism could help explain why European winters were so severe during the Maunder Minimum. During periods of very low solar activity (with fewer sunspots), more galactic cosmic rays enter the atmosphere. This would have been true during the Maunder Minimum. These cosmic rays increased the number of ions in the atmosphere. Normally, decreases in the voltage between the capacitor plates or other factors could overpower or obscure this effect. However, at times of very low solar activity, the number of cosmic rays entering the atmosphere is so great that this single effect becomes more important than changes in the other four factors. So, we expect J_z to be consistently larger at times of very low solar activity, in spite of the fact that the other four factors can change. Also, for reasons I explain in the technical articles, we expect this effect to be greatest during winter in the northern high latitudes.[39] This increased J_z caused winter cyclones in the northern Atlantic to become more intense, accompanied by more intense blocking events to the east. These blocking events prevented the west-to-east flow of air, which prevented warm, moist air from the Atlantic from reaching western Europe. Because water vapor is an effective greenhouse gas, this lack of water vapor over the European continent led to more energy being radiated to outer space, resulting in much colder and more severe European winters.

During the coldest part of the Little Ice Age, which was also a time of very low sunspot number, the Thames River in England froze, allowing "frost fairs" to be held on the ice

It is worth noting that the European winters of 2009–2011 were quite severe, and those were also periods of very low solar activity (few sunspots). Hence, if this mechanism is correct, then we should see additional severe winters in Europe during future periods of very low solar activity.

Many have speculated that solar activity could also be affecting global climate, not just climate in the northern Atlantic region. This is possible, but one would probably need to include this effect in sophisticated computer models in order to determine whether this is the case or not.

I would be remiss if I did not point out that this proposed mechanism is Dr. Tinsley's, not mine. Furthermore, Dr. Tinsley does not share my creationist views. Also, his views on global warming are pretty conventional in that he thinks man-made global warming is real, although he doesn't think it's a cause for alarmism. Although I have tried to keep an open mind, I am becoming increasingly skeptical of this, especially since much of the evidence for global warming (especially catastrophic global warming) depends on dubious interpretations of past climate data and uncertain parameters in climate models. In any case, as I hope this book has made clear, I too do not think that climate change is something that people should be panicking over.

References

1. Herschel, W. 1801. Observations tending to investigate the nature of the Sun, in order to find the causes or symptoms of its variable of light and heat; with remarks on the use that may possibly be drawn from solar observations. *Philosophical Transactions of the Royal Society of London*. 91: 265-318.
2. Carslaw, K. S., R. G. Harrison, and J. Kirkby. 2002. Cosmic Rays, Clouds, and Climate. *Science.* 298: 1732-1737.
3. Hargreaves, J. K. 1995. *The Solar-Terrestrial Environment*. Cambridge: Cambridge University Press, 141.
4. Dickinson, R. E. 1975. Solar variability and the lower atmosphere. *Bulletin of the American Meteorological Society*. 56 (12): 1240-1248.
5. Carslaw, Cosmic Rays, Clouds, and Climate, 1734-1736.
6. Vardiman, L. 2008. A New Theory of Climate Change. *Acts & Facts.* 37 (11): 10.
7. Vardiman, L. 2011. CLOUD Experiment Supports Global Warming Theory. *Acts & Facts.* 40 (11): 22-23.
8. Rogers, R. R. and M. K. Yau. 1996. *A Short Course in Cloud Physics*, 3rd ed. Burlington, MA: Butterworth-Heinemann, 81.
9. Svensmark, H. and E. Friis-Christensen. 1997. Variation of cosmic ray flux and global cloud coverage—a missing link in solar-climate relationships. *Journal of Atmospheric and Solar-Terrestrial Physics*. 59 (11): 1225-1232.
10. Marsh, N. D. and H. Svensmark. 2000. Low Cloud Properties Influenced by Cosmic Rays. *Physical Review Letters*. 85 (23): 5004-5007.
11. Marsh, N. and H. Svensmark. 2003. Galactic cosmic ray and El Niño-Southern Oscillation trends in International Satellite

Cloud Climatology Project D2 low-cloud properties. *Journal of Geophysical Research.* 108 (D6): 4195.

12. Kirkby, J. et al. 2011. Role of sulphuric acid, ammonia and galactic cosmic rays in atmospheric aerosol nucleation. *Nature.* 476: 429-433.
13. Vardiman, CLOUD Experiment Supports Global Warming Theory.
14. Kirkby et al, Role of sulphuric acid, ammonia and galactic cosmic rays in atmospheric aerosol nucleation.
15. Pierce, J. R. and P. J. Adams. 2009. Can cosmic rays affect cloud condensation nuclei by altering new particle formation rates? *Geophysical Research Letters.* 36: L09820.
16. Svensmark, H., M. B. Enghoff, and J. O. P. Pedersen. 2013. Response of Cloud Condensation Nuclei (> 50 nm) to changes in ion-nucleation. *Physics Letters A.* 377 (37): 2343-2347.
17. Rogers and Yau, *A Short Course in Cloud Physics,* 124.
18. Tinsley, B. A. 2010. Electric charge modulation of aerosol scavenging in clouds: Rate coefficients with Monte Carlo simulation of diffusion. *Journal of Geophysical Research.* 115 (D23): D23211.
19. Pruppacher, H. R. and J. D. Klett. 1997. *Microphysics of Clouds and Precipitation*, 2nd ed. Dordrecht, Netherlands: Kluwer Academic, 711.
20. Mason, B. J. 1971. *The Physics of Clouds*, 2nd ed. Oxford: Clarendon Press, p. 63.
21. Zhou, L., B. A. Tinsley, and A. Plemmons. 2009. Scavenging in weakly electrified saturated and subsaturated clouds, treating aerosol particles and droplets as conducting spheres. *Journal of Geophysical Research.* 114: D18201.
22. Another factor called the curvature effect also comes into play, but we will not get into that here. See the curvature effect discussed in Pruppacher and Klett, *Microphysics of Clouds and Precipitation.*
23. Bering, E. A. III, A. A. Few, and J. R. Benbrook. 1998. The Global Electric Circuit. *Physics Today.* 51 (10): 24-30.
24. Tinsley, B. A., G. B. Burns, and L. Zhou. 2007. The role of the global electric circuit in solar and internal forcing of clouds and climate. *Advances in Space Research.* 40 (7): 1126-1139.
25. Pruppacher and Klett, *Microphysics of Clouds and Precipitation*, 802.
26. Tinsley et al, The role of the global electric circuit in solar and internal forcing of clouds and climate.
27. Hebert, J. 2013. Two possible mechanisms linking cosmic rays to weather and climate. *Journal of Creation.* 27 (2): 91-98.
28. Hebert, J. 2013. Apparent difficulties with a CMAS cosmic ray-weather/climate link. *Journal of Creation.* 27 (3): 93-97.
29. Hebert, J. 2014. Are cosmic rays affecting high-latitude winter cyclones? *Journal of Creation.* 28 (1): 59-67.
30. Hebert, J. 2014. Solar activity, cold European winters, and the Little Ice Age. *Journal of Creation.* 28 (1): 114-121.
31. Hebert, L. 2011. Atmospheric Electricity Data from Mauna Loa Observatory: Additional Support for a Global Electric Circuit-Weather Connection? Ph.D. dissertation, University of Texas at Dallas.
32. Hebert, L., B. A. Tinsley, and L. Zhou. 2012. Global electric circuit modulation of winter cyclone vorticity in the northern high latitudes. *Advances in Space Research.* 50 (6): 806-818.
33. Hebert, L., B. A. Tinsley, and L. Zhou. 2011. Is the global Electric Circuit Modulating Winter Cyclone Vorticity in the Northern high Latitudes? American Geophysical Union poster abstract. Archived at http://adsabs.harvard.edu/abs/2011AGUFMAE31A0255H, accessed November 30, 2018.
34. The Sun's Chilly Impact on Earth. NASA. Posted on giss.nasa.gov December 6, 2001, accessed August 13, 2019.
35. Wiedenmann, J. M. et al. Tikhonova. 2002. The climatology of blocking anticyclones for the northern and southern hemispheres: Block intensity as a diagnostic. *Journal of Climate.* 15: 3459-3473.
36. Lupo, A. R. and P. J. Smith. 1994. Climatological features of blocking anticyclones in the Northern Hemisphere. *Tellus.* 47A: 439-456.
37. Lutgens, F. K., E. J. Tarbuck, and D. Tasa. 2010. *The Atmosphere: An Introduction to Meteorology*, 11th ed. New York: Prentice Hall, 258.
38. Barriopedro, D., R. García-Herrera, and R. Huth. 2008. Solar modulation of Northern Hemisphere winter blocking. *Journal of Geophysical Research.* 113: D14118.
39. Hebert, Are cosmic rays affecting high-latitude winter cyclones?, 62-63.

B Appendix

Stacking and Tuning: Paleoclimate Speculation Explained in Four Easy Steps

Summary: Scientists attempt to infer a history of past climate change by analyzing chemical clues within seafloor sediments and ice cores. When measured quantities from the cores are plotted on graphs, many wiggly patterns become apparent. The wiggles from different sediment and ice cores seem to tell a consistent climate change "story," one that appears to be in very good agreement with Milankovitch expectations. At first glance, this might seem to be a very strong argument for the Milankovitch theory. However, the apparent agreement does not confirm the theory. Rather, this agreement is obtained by assuming the theory is correct and then using it to assign ages to these chemical wiggles in a process called orbital tuning. This process requires some parts of a signal to be compressed and others stretched, much like an accordion. Orbital tuning provides an enormous opportunity for self-deception. In one study, a scientist convincingly tuned a randomly generated computer signal to Milankovitch expectations—even though the signal had absolutely nothing to do with the real world. For these reasons, the apparent agreement between different climate data sets, though superficially impressive, cannot be used as evidence that the Milankovitch theory is correct.

Have you ever wondered how uniformitarian scientists think they can infer details about local or regional climate changes that supposedly occurred in the distant past? The purpose of this appendix is to pull back the curtain and remove some of the mystery on this subject. I hope to explain in simple terms how uniformitarian scientists attempt to infer information about past climates, and I want to explain why those inferences are wrong.

We begin our discussion with Figure B.1. Technical papers, as well as popular-level articles on the internet, often show figures such as this one. Climate variables, deduced from oxygen isotope values in seafloor sediments and ice cores, seem to track nearly perfectly with one another for hundreds of thousands of years in a pronounced sawtooth pattern. Moreover, they seem to be in perfect agreement with Milankovitch expectations. At first glance, this seems to be an incredibly powerful argument for an old earth as well as the Milankovitch theory. After all, if multiple independent data sets from all over the world are telling a consistent story about climate change, a story that goes back into the distant prehistoric past, then how can any reasonable person doubt that the earth is truly ancient? And who can doubt that the Milankovitch theory is correct? I recall seeing these kinds of graphs in graduate school. Even though I was a firmly convinced and well-informed creationist, I didn't know how to explain the apparent agreement between these climate variables and the Milankovitch theory. I had to admit it looked very convincing.

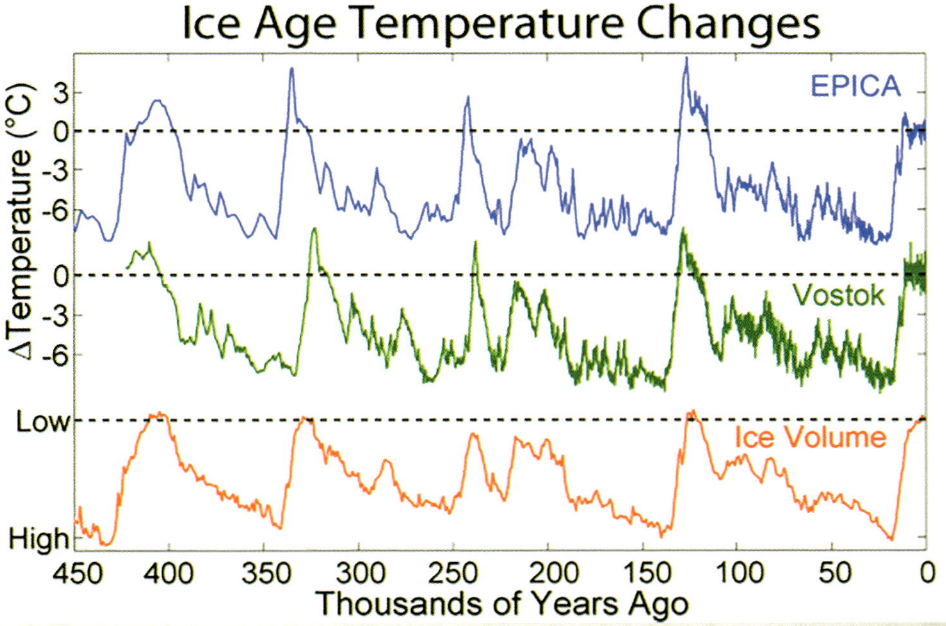

Figure B.1. Graphs of climate variables versus secular age assignments are common in both the technical and popular literature. These graphs seem to show a consistent story of climate change going back hundreds of thousands of years. The consistent, synchronized behavior of these graphs seems, at first glance, to be a powerful argument for both an old earth and the Milankovitch Ice Age theory. However, this agreement is obtained by assuming the Milankovitch theory is correct and using that theory to assign ages to the climate variable wiggles. This process is called orbital tuning.

These kinds of graphs also show atmospheric carbon dioxide tracking with inferred temperatures and other climate variables. Commenting on a similar comparison of temperature, atmospheric carbon dioxide, and global average sea level, one climate researcher stated:

> The correlation between these completely independent datasets, one from Antarctic cores, the other from deep sea foraminifera, is astounding. It demonstrates that climate and the carbon cycle are tightly interlinked. High CO_2 concentrations are always associated with warm temperatures, high sea level, and low ice volume. This indicates the importance of atmospheric CO_2 concentrations for climate but it also suggests that climate impacts the carbon cycle and causes changes in CO_2.[1]

However, once you understand the concept of orbital tuning, the apparent agreement between the data sets in Figure B.1 is not nearly as impressive as it initially seems. The big problem is that contrary to their appearances, these data sets are not truly independent.

Variable Sedimentation Rates

Although uniformitarian scientists believe seafloor sediments were deposited slowly and gradually on the ocean floor, they recognize that these rates of deposition will not be perfectly constant. Sometimes sediments will accumulate on the ocean floor a little faster, and at other times they will accumulate a little slower. But changes in how fast sediments are settling on the ocean floor complicates the analysis of those chemical wiggles.

As an example, consider a device that is used to measure earthquake activity: a seismograph. Seismographs spool out paper at a constant rate, and a stylus indicates earthquake activity as wiggles on the paper.

Suppose we had such a device, but instead of measuring earthquake activity, the stylus was programmed to move up and down in a uniform, rhythmic motion. Suppose also that the paper was being spooled out at a perfectly constant rate. In that case, a simple wavelike pattern would be produced (Figure B.2). But suppose the spool rolled out the paper at a variable rate instead. Sometimes it spooled out the paper faster than average, and at other times it spooled out slower than average. In that case,

A seismograph is used to record earthquake activity

a pattern like that in Figure B.3 would be produced. The wiggles on the paper will be narrower when the paper is moving more slowly, and the wiggles will be broader when the paper is moving more quickly.

Something similar happens when sediments accumulate at a variable rate on the ocean floor. If sediments accumulated on the seafloor at a perfectly constant rate, the apparent widths of oxygen isotope wiggles would be completely undistorted.[2] A plot of oxygen isotope wiggles as a function of depth would also be, in effect, a plot of those same wiggles as a function of time. If a graph of oxygen isotope values versus core depth showed a simple wavelike pattern like that in Figure B.2, then a graph of those same values versus age would also show that same simple wavelike pattern. On the

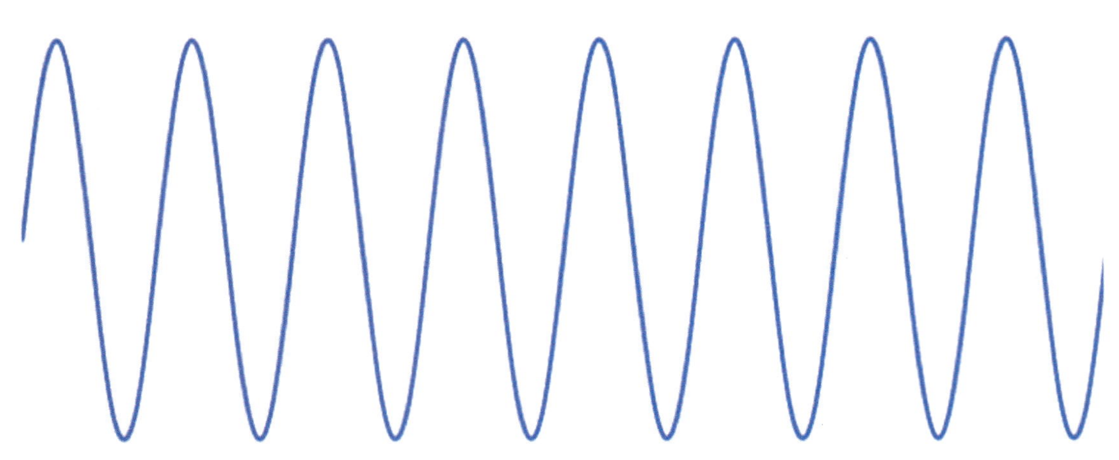

Figure B.2. A rhythmic up-and-down motion of the stylus will produce a simple wave pattern if the paper is spooled out at a constant rate

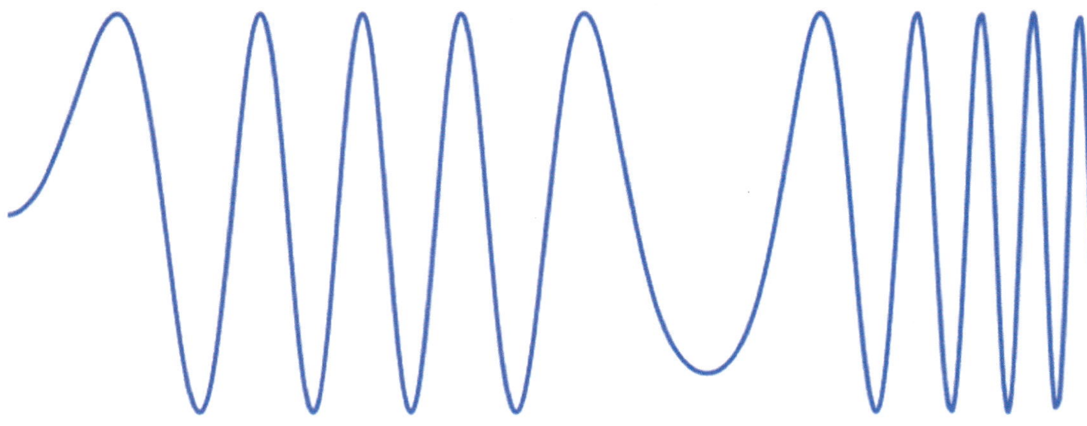

Figure B.3. If the rate at which the paper is spooled out varies, some of the wiggles will be stretched out and others will be compressed. This distortion is similar to the way that changing sedimentation rates distorts an oxygen isotope signal in seafloor sediments.

other hand, if the rate of deposition varied in the past so that it was a little faster or slower at times than on average, then the apparent width of the wiggles will be distorted like we see in Figure B.3. At times when the sediments were deposited more slowly, the spacing between wiggles will be narrower. Likewise, at times when the sediments were deposited more quickly, the spacing between wiggles will be wider.

Step 1: Alignment

Remember that uniformitarian scientists think that oxygen isotope ($\delta^{18}O$) values in seafloor sediments indicate how much ice was on Earth at different times in the past. The highest oxygen isotope values represent the deepest parts of past Ice Ages, when global ice volume was supposedly greatest. Likewise, the lowest oxygen isotope values represent the warmest parts of the warmer periods (interglacials), when global ice volume was at a minimum. They also think the oxygen isotope wiggles should be a global climate indicator—at least in theory. This means that a $\delta^{18}O$ wiggly pattern from one core should ideally be telling the same story as the $\delta^{18}O$ wiggles from another core.

However, as discussed in chapter 17, the $\delta^{18}O$ wiggles in a core are also influenced by local changes in temperature and seawater chemistry. This means that the $\delta^{18}O$ wiggles in a sediment core are not a "pure" indicator of global ice volume. Some of the variation in $\delta^{18}O$ values in a core is due to these local effects. Nevertheless, secular scientists think that most of the variation in the $\delta^{18}O$ values is due to changes in global ice volume.

If that's true, then a prominent $\delta^{18}O$ peak in one core should be the same age as a corresponding prominent $\delta^{18}O$ peak in another core. Likewise, a prominent trough in one core should be the same age as a corresponding prominent trough in another core. However, particular $\delta^{18}O$ features in one core are almost always at different depths than the presumably corresponding $\delta^{18}O$ features in another core (Figure B.4). Uniformitarian scientists attribute this to the fact that sedimentation rates can vary from core to core and vary over time.

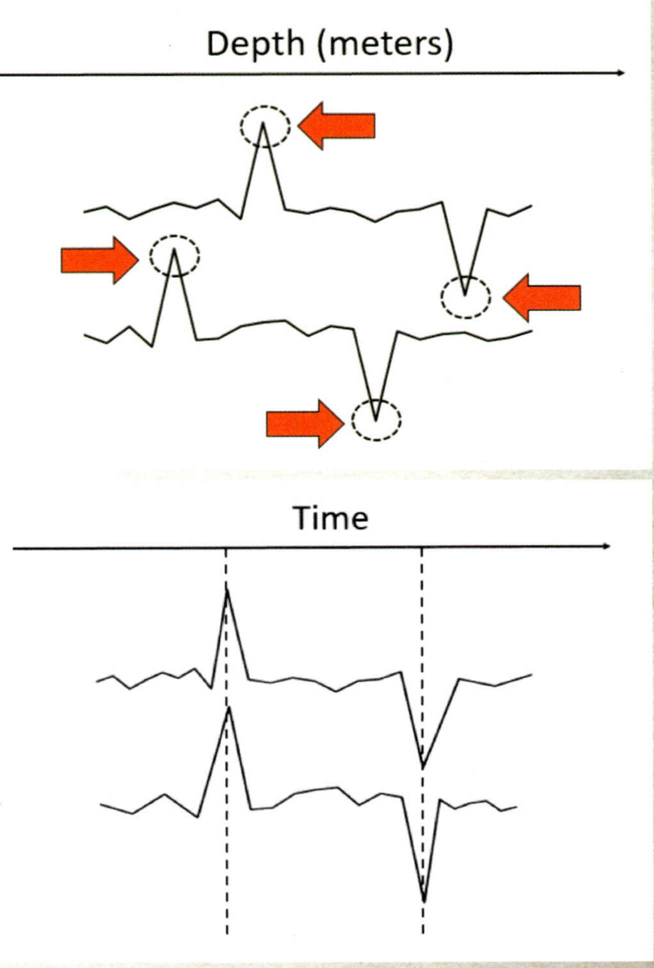

Figure B.4. Secular paleoclimatologists align similar-looking oxygen isotope features in different sediment cores. This alignment process requires some parts of an oxygen isotope record to be compressed and others stretched in an accordion-like fashion. After alignment, the orbital tuning technique is used to assign ages to the wiggle pattern.

"They are simply assuming that similar-looking features in two different $\delta^{18}O$ signals are the same age even though we really don't know that to be the case."

Since uniformitarian scientists think these corresponding peaks are the same age, they will align the peaks and troughs in one core with the peaks and troughs in another core (Figure B.4). In doing so, the different $\delta^{18}O$ signals are placed on a common, but unspecified, timescale. At this point, the actual numerical times aren't important. The important thing is just to make sure that the similar $\delta^{18}O$ features have been aligned with one another.

Of course, this alignment process means that some parts of an oxygen isotope signal will be stretched and other parts will be squished in an accordion-like fashion. However, it's important to realize that secular scientists don't think they are cheating. In their minds, they are simply removing distortions caused by changes in past sedimentation rates. They see this as part of the process of restoring the original *undistorted* $\delta^{18}O$ signals. In other words, they are attempting to recover what these $\delta^{18}O$ signals are supposed to look like had sedimentation rates not varied over time.

You can probably already see some problems with this. First, none of us was there to actually observe the deposition of these sediments. This means we don't know how fast the sediments were deposited. That, in turn, means that we don't know what the signal is supposed to look like. Second, they are simply assuming that similar-looking features in two different $\delta^{18}O$ signals are the same age even though we really don't know that to be the case. This is because other local factors can also affect the values of these $\delta^{18}O$ values (see below).

Step 2: Stacking

Secular scientists recognize that local factors such as changes in ocean chemistry or temperature *also* influence the $\delta^{18}O$ values in the sediments. These local effects will cause some $\delta^{18}O$ values to be a little higher than they should be and other values to be lower than they should be.

Is there a way to theoretically remove these local influences and leave behind just a pure $\delta^{18}O$ signal, one that more accurately records global ice volume over time? Yes, it's called *stacking*.[3] In stacking, prominent $\delta^{18}O$ features from different cores are first aligned (as in Figure B.4), and then they are averaged together. The thinking is that a local distortion in a particular $\delta^{18}O$ value is just as likely to be positive as it is to be negative. If a large number of these $\delta^{18}O$ signals are averaged together, then these positive and negative distortions should cancel out, leaving just a pure global $\delta^{18}O$ signal that is a true record of global ice volume.

Two polar scientists in red polar clothing working on ice cores over an ice floe

Of course, there are a few complications. The average $\delta^{18}O$ values from one particular foraminifera can be higher or lower than the average $\delta^{18}O$ values from another species of foram. So, the person doing the stacking will often have to add or subtract a constant value from some of the data sets to make sure that all the separate data sets have the same average value. Also, $\delta^{18}O$ peaks in one core may generally be taller than those in another core. The person doing the stacking may need to scale the $\delta^{18}O$ peaks in each core so that the highest peaks in each core are about the same height. But aside from those complications, stacking is just the process of averaging the $\delta^{18}O$ values from different cores in an attempt to remove local noise.

Step 3: Tuning

Orbital tuning uses the Milankovitch theory to assign ages to $\delta^{18}O$ wiggles in either an individual core or a stack of $\delta^{18}O$ values. Remember that secular scientists think the Milankovitch theory tells them the approximate times at which the Ice Ages occurred. These times are approximate because scientists still have to make assumptions about how long it takes the climate to respond to the changes in sunlight. These details are filled in with a model of some kind. One popular model is an equation that gives the rate of change of global ice volume as a function of the amount of summer sunlight striking the high latitudes and the ice volume itself.[4] This model has several adjustable parameters. Scientists choose values for these parameters that best fit the oxygen isotope data.

This model assumes that thick ice sheets build up slowly but collapse quickly. The result is the recognizable "sawtooth" pattern such as that shown in Figure B.1. In fact, if you read the bottom-most graph in Figure B.1 from left to right, the wiggles are telling a story about how ice volume has varied over time. Note that this graph is *upside down* so that low volumes of ice are on top and high values of ice are on bottom. The ice takes tens of thousands of years to build up to a maximum value, but then it quickly drops to minimum values. Of course, this should not be too surprising because the model assumes this will happen.

The Data Sets Are Not Independent

Before explaining step 4, we are now in a position to explain why the apparent agreement between the different wiggles in Figure B.1 is not the slam-dunk argument for secular climate change stories that it superficially appears to be. The wiggles in these three graphs align well, but this good alignment became apparent after they were

“These wiggles align because they have been *made* to align. Even if the temperature and ice volume values were truly independent of one another, the ages assigned to these values are not independent.”

placed on a common timescale. Remember, these values came from quantities that were measured in cores as a function of depth, not time. So, secular scientists somehow had to transform these depths into ages. Where did these ages come from? They ultimately came from the Milankovitch theory! Secular scientists use the Milankovitch theory and orbital tuning to assign ages to $\delta^{18}O$ wiggles in deep-sea sediments. Then they transfer these ages to similar-looking wiggles in ice cores. These wiggles align because they have been *made* to align. Even if the temperature and ice volume values were truly independent of one another, the ages assigned to these values are not independent.

Even two randomly generated signals will have similarities between them. Some peaks and troughs will be more conspicuous than others. If you are given the freedom to selectively stretch and compress different parts of the two signals, there is a good chance that you can force the kind of agreement that is seen in Figure B.1. That is precisely what orbital tuning does—it selectively stretches and compresses different parts of two climate signals to force them into an apparent alignment. This is why creation scientists do not consider these graphs to be convincing proof of an old earth or the Milankovitch theory.

Step 4: Comparison

The final step is to compare the single δ¹⁸O record with the stacked record and to explain why they sometimes differ from one another. After tuning, there will generally be overall good agreement between the stacked record and the test record (Figure B.5). However, the agreement will not be perfect. Sometimes the δ¹⁸O values in the test core will be higher than those in the stack. At other times, they will be lower.

Remember that uniformitarian scientists think that the stacking process removed local effects from their stacked δ¹⁸O signal. So, if the test signal somehow differs from the stack, uniformitarian scientists will probably attribute this to local effects at the test core location. For instance, they could argue that perhaps the ocean circulation changed dramatically at the core location at that time, or maybe the ocean chemistry changed—or some other such thing. This is how uniformitarian scientists think they can infer particular changes in climate (even changes in

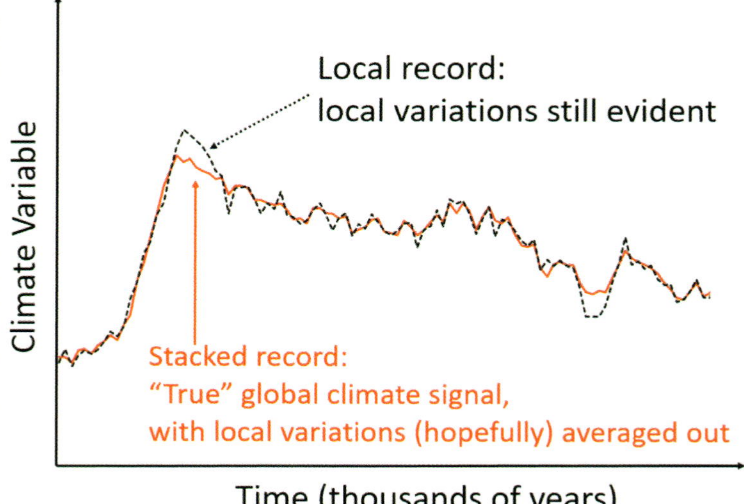

Figure B.5. If a tuned climate signal from one particular sediment core differs somewhat from a tuned and stacked signal (which is thought to represent a true global climate signal), differences between the two signals can be attributed to local factors at the individual core site. Stacked record construction using data from reference 3.

local climate) that supposedly occurred in the distant past.

Overtuning

Could secular scientists simply be fooling themselves when they orbitally tune chemical wiggles from the seafloor sediments or ice cores? Creation scientists think so. Remember that any wiggly pattern will have peaks and troughs, and some of those peaks will necessarily be more prominent than others. Likewise, some troughs will be deeper than others. If one tries hard enough, peaks and troughs in different signals can usually be aligned in a fairly plausible way, even if the two signals are totally unrelated to one another. Even secular scientists have acknowledged this possibility.[5-7] In one study, a scientist convincingly tuned a randomly generated computer signal to Milankovitch expectations—even though the signal had absolutely nothing to do with the real world![7]

Now that you know this, it should be clear why graphs such as Figure B.1 are not a strong argument for an old earth or the Milankovitch theory. Secular scientists use a Milankovitch-based model to assign ages to chemical wiggles in sediment cores, so the ages assigned to these different data sets are not independent. Hence, the apparent agreement between different data sets, as well as their apparent agreement with Milankovitch expectations, is hardly surprising.

References

1. Schmittner, A. Introduction to Climate Science. Online text for Oregon State University, College of Earth, Ocean, and Atmospheric Sciences. Posted on library.open.oregonstate.edu, accessed January 4, 2019.
2. For the simplicity of this discussion, I am ignoring distortions that occur after the sediments were deposited. Sometimes marine organisms can disturb the sediments (bioturbation), and sometimes ocean-bottom currents can disturb them as well.
3. Lisiecki, L. E. and M. E. Raymo. 2005. A Pliocene-Pleistocene stack of 57 globally distributed benthic $\delta^{18}O$ records. *Paleoceanography and Paleoclimatology.* 20: PA1003.
4. Imbrie, J. and J. Z. Imbrie. 1980. Modeling the Climatic Response to Orbital Variations. *Science.* 207: 943-953.
5. Blaauw, M. 2012. Out of tune: the dangers of aligning proxy archives. *Quaternary Science Reviews.* 36 (12): 38-49.
6. Blaauw, M., K. D. Bennett, and J. A. Christen. 2010. Random walk simulations of fossil proxy data. *The Holocene.* 20 (4): 645-649.
7. Neeman, B. U. 1993. Orbital Tuning of Paleoclimatic Records: A Reassessment. Ernest Orlando Lawrence Berkeley National Laboratory. LBNL-39572, UC-412.

C Appendix

More Problems with the Pacemaker Paper

Summary: According to the Milankovitch Ice Age theory, slight changes in Earth's orbital and rotational motions cause changes in the distribution of sunlight falling on Earth, and these changes in solar radiation control the timing of Ice Ages. The modern version of this theory is widely accepted because an iconic paper titled "The Pacemaker of the Ice Ages" published in 1976 seemed to provide evidence for the theory. Data from two deep-sea sediment cores seemed to agree well with Milankovitch expectations. However, this paper's confirmation of the theory has serious issues. Its results were critically dependent upon an age assignment of 700,000 years for the most recent reversal of the earth's magnetic field, but this age assignment was revised years later, significantly weakening, if not completely invalidating, evidence for the theory. Additionally, the paper's authors excluded nearly a third of the data from the second core, claiming that the uppermost section of that core might have been too old to date with radiocarbon dating. Yet, they did not even attempt to radiocarbon-date the top of the core. This exclusion was problematic because in a shorter data set it is more likely that random wiggle patterns in the data might agree with Milankovitch expectations simply by chance. So, their exclusion of a third of the data from the second core made it much easier for them to obtain results from the core that favored the theory. Even stranger, the authors did not plot oxygen isotope data from the top of the second core in their paper's figures, even though other data sets from the top of that core were plotted. This omission made it harder for others to judge whether this exclusion of data was reasonable. Also, the original, unaltered data used in the analysis do not seem to be publicly available, which is very strange in light of the paper's importance.

In chapter 9, I described how the "Pacemaker of the Ice Ages"[1] paper published in the journal *Science* in 1976 convinced many secular scientists that the Milankovitch theory is correct, in spite of many well-known problems with the theory. That paper's results were critically dependent upon an age assignment that secular scientists themselves no longer accept as valid. Redoing these calculations after taking this age revision into account significantly weakens, if not completely invalidates, this iconic argument for the Milankovitch theory. Furthermore, the paper had other serious problems. In this appendix, I explain these problems in more detail for interested readers.

The Pacemaker paper presented an argument for the Milankovitch theory using data from two deep-sea sediment cores from the far southern Indian Ocean, designated as RC11-120 and E49-18. Another core from the western Pacific designated as V28-238 also played an important role in the analysis (Figure C.1).

The Pacemaker authors analyzed three different data sets. The first data set was composed of oxygen isotope ($\delta^{18}O$) measurements from the shells of microscopic creatures called *planktonic foraminifera* within the sediments. Of the three data sets they analyzed, this oxygen isotope ($\delta^{18}O$) data set is arguably the most important. They also inferred summer sea-surface temperatures in the southern hemisphere from the shells of other microscopic creatures called *radiolarians* that were within the sediments, and they used the percent abundance of one particular radiolarian species as a third climate indicator.

The authors used spectral analysis to search for prominent cycles (presumably climate cycles) within the data. They found cycles having lengths of about 100K, 42K,

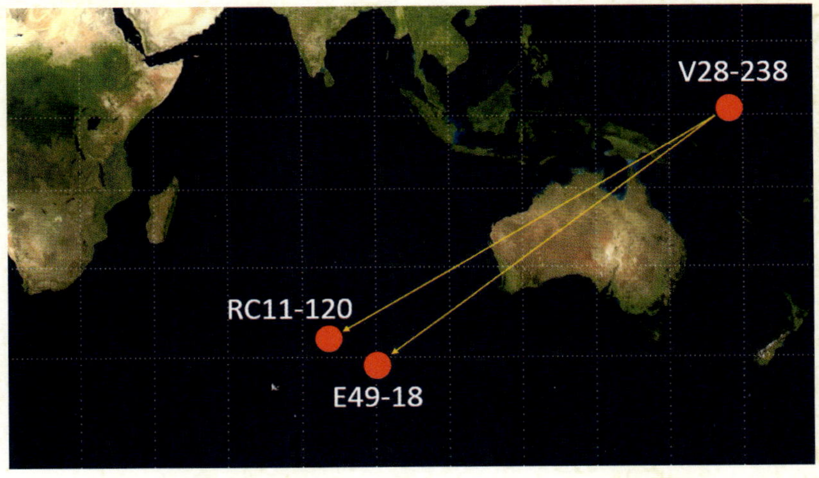

Figure C.1. Chemical wiggles in the far western Pacific core V28-238 were used to assign ages to chemical wiggles in the RC11-120 and E49-18 sediment cores that were used in the famous "Pacemaker of the Ice Ages" paper.

and 23K years. They also detected a cycle length of 19,000 years, although other uniformitarian scientists have since argued that this apparent cycle was not real.[2] Because these cycle lengths were very close to the lengths of 100K, 41K, and 23K years resulting from astronomical calculations, secular scientists saw this as strong evidence for the Milankovitch theory.

However, there are serious problems with this paper.[3-8] The most obvious problem is that the Pacemaker results depended upon an assumed age of 700,000 years for the most recent flip or reversal of the earth's magnetic field, even though secular scientists now claim this event happened 780,000 years ago. This change by itself is sufficient to call the Pacemaker results into question. But the paper also had other serious problems.

The Pacemaker authors excluded nearly one-third of the data in the longer E49-18 sediment core from their analysis. They suggested that currents on the ocean bottom had perhaps scoured away the top portion of the core. They also suggested that perhaps the E49-18 core top was as old as 60,000 years, which would make it too old to date with the radiocarbon dating method. Since they supposedly could not assign a date to the core top, this meant that the top section of the core could not be dated, making it unusable for purposes of their analysis. Hence, they excluded data from the upper third of this core from their study.

However, other uniformitarian scientists later disagreed with this conclusion, arguing that the core top was just 12,000 years old.[9] In that case, the core top was potentially datable with the radiocarbon method, and data from the upper core portion could have, and *should* have, been used in their analysis. Yet, the Pacemaker authors did not even attempt to radiocarbon date the top of the E49-18 core, even though they used a radiocarbon date from the shorter RC11-120 core in their Pacemaker analysis. Had they done so, one of two things would have happened: Either the amount of radiocarbon detected would have been too low to obtain a precise radiocarbon date, which would have justified their exclusion of this section of data, or the amount of detected radiocarbon would have been large enough to allow precise dating of the core top.

Why did they make no attempt to date the top of the E49-18 core? I strongly suspect it is because, deep down, they didn't really trust radiocarbon dating to give them the answer they wanted. They were probably afraid that a radiocarbon date for the top of the core would be young. In that case, they would have been obligated to include the uppermost E49-18 data in their analysis. This greater amount of data would have greatly increased the likelihood that the results from spectral analysis would not agree with Milankovitch expectations, as I explain below.

“Why did they make no attempt to date the top of the E49-18 core? I strongly suspect it is because, deep down, they didn't really trust radiocarbon dating to give them the answer they wanted.”

This apparent concern was well-justified. I redid the analysis, taking into account those data and an assumed age of 12,000 years for the top of the E49-18 core. I used two different age models, the first based on just two age control points, and the second based on three such points. In the first case, the results were wildly inconsistent with Milankovitch expectations. In the second case, two out of the three variables were reasonably consistent with those expectations, but the $\delta^{18}O$ spectral results (arguably the most important) were not.[10]

"It is also worth noting that this radiocarbon date was the *only* age assignment in the Pacemaker paper based on a direct dating of sediments within the two Indian Ocean sediment cores."

By the way, the radiocarbon date for the shorter RC11-120 core they used in the analysis was not an important age control point, and they could have easily excluded it from the analysis without affecting the overall results. It is also worth noting that this radiocarbon date was the *only* age assignment in the Pacemaker paper based on a direct dating of sediments within the two Indian Ocean sediment cores. All the other age assignments were based on radioisotope dates from other locations that were then transferred to these two sediment cores.

So, why is the exclusion of the uppermost data from the E49-18 core a problem? The default assumption (null hypothesis) when analyzing chemical wiggles within the seafloor sediments is that they were generated by a random process. Suppose, however, that one is hoping that the wiggles instead were generated by some nonrandom process, such as some kind of astronomical climate forcing. In order to make a convincing case for the nonrandom process, the wiggles have to behave in a certain way for some minimum amount of time. However, if the wiggles were instead generated by a random process, then a greater number of observed wiggles would make their random nature more evident.

As an example, consider the process of flipping a coin. Suppose you're trying to determine whether or not the coin is fair. In this case, the null hypothesis (default assumption until shown otherwise) is that it's fair, with a 50/50 probability of landing either heads or tails. In order to disprove your default assumption, you would need to get quite a few heads or tails in a row, too many to reasonably account for by chance. Would getting two or three heads in a row be sufficient to falsify your null hypothesis? No, because getting just two or three heads in a row is not that hard to do. However, getting five or six heads in a row is much harder to do with a fair coin. Likewise, the fact that chemical wiggles within a short section of sediment core by chance behave in accordance with Milankovitch expectations is not terribly impressive. In order to make a strong case for the Milankovitch hypothesis, the wiggles within a *long* core section need to behave in accordance with the theory.

This is why the exclusion of nearly one-third of the data from the second sediment core was problematic. In excluding these data, the Pacemaker authors made it easier to obtain results from the E49-18 core that were consistent with Milankovitch expectations. As noted earlier, inclusion of all the E49-18 data with an assumed age of 12,000 years for the core top undermines the argument for the Milankovitch theory. Of course, this age estimate of 12,000 years was not based on anything firm—the scientists who provided it did not give any explanation for their age assignment.[11] It seems to have been a simple guess based on the shape of the $\delta^{18}O$ wiggles at the top of the core.

But this leads to another serious problem with the Pacemaker paper. The reason secular scientists think they can estimate ages from just the shape of $\delta^{18}O$ wiggles is that they have a preconceived idea of what oxygen isotope wiggles of a certain age are supposed to look like. However, not only did the Pacemaker authors not use the data from the top of the E49-18 core, including the uppermost $\delta^{18}O$ data, but they did not even plot the uppermost $\delta^{18}O$ data on their Figure 3. This is odd, especially since they plotted values from three other data sets on that part of their graph.

Had the Pacemaker authors actually plotted these uppermost E49-18 $\delta^{18}O$ wiggles on their graph, other secular scientists might have concluded, based on the shape of those $\delta^{18}O$ wiggles, that the top of the E49-18 core was quite young. This can be seen by comparing the oxygen isotope wiggles at the tops of both the RC11-120 and E49-18 cores (Figure C.2). The blue rectangle shows the oxygen isotope data the Pacemaker authors did *not* plot on their Figure 3. However, if they had plotted those uppermost E49-18 $\delta^{18}O$ values on their Figure 3, then those $\delta^{18}O$ values would have trended upward, as do the $\delta^{18}O$ values at the top of the RC11-120 core. A secular paleoclimatologist might easily have concluded that the very uppermost E49-18 $\delta^{18}O$ values (indicat-

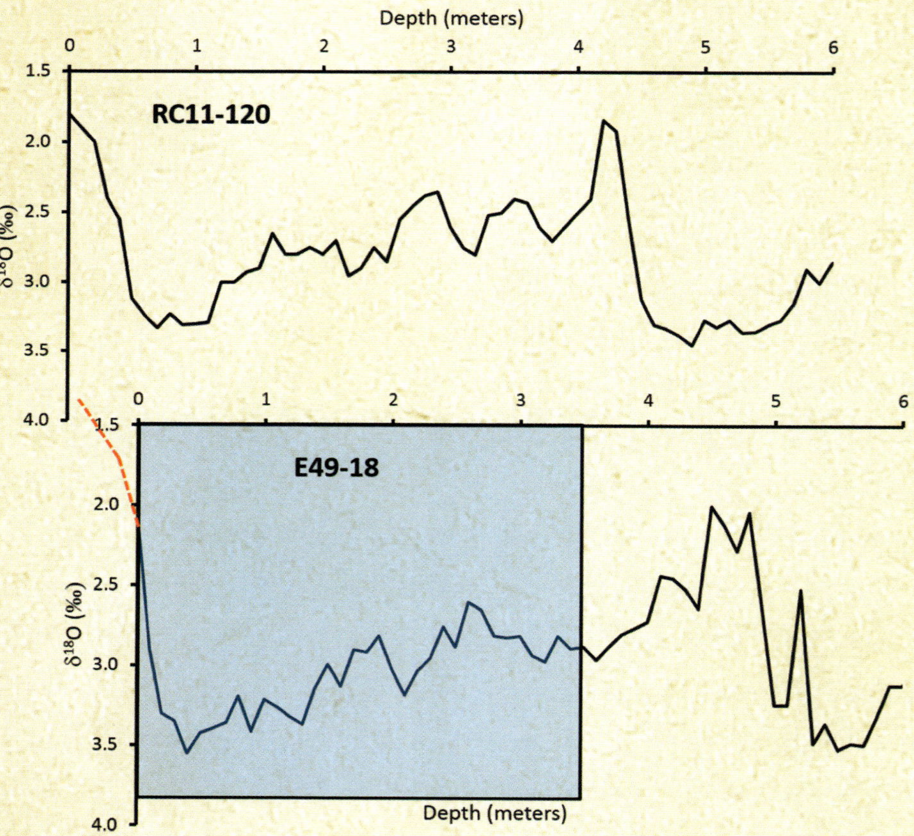

Figure C.2. Had the Pacemaker authors plotted the oxygen isotope ($\delta^{18}O$) data from the upper section of the longer E49-18 core (highlighted in blue) on their Figure 3, other secular scientists might have concluded, based on the shape of those uppermost $\delta^{18}O$ wiggles, that this part of the E49-18 core should have been included in the Pacemaker analysis. Inclusion of this section of data would likely have resulted in results that were unfavorable to the Milankovitch theory.

ed by the dashed red lines) had indeed been scoured away for some reason. But given the similarity in the upward trends shared by both graphs, he would likely conclude that the uppermost remaining $\delta^{18}O$ values in the E49-18 core were still quite young and that they corresponded to the $\delta^{18}O$ values, which were also trending upward, within the uppermost meter of the RC11-120 core.

If the uppermost E49-18 $\delta^{18}O$ values really were young, then secular scientists should have attempted to date them by radiocarbon dating. In fact, had the Pacemaker authors plotted those uppermost E49-18 $\delta^{18}O$ values on their Figure 3, it is quite likely that one of the papers' referees would have demanded that they perform such a radiocarbon test. This radiocarbon date would likely have yielded a very young age for those uppermost E49-18 sediments. Using sediments in the analysis would have likely yielded results that were unfavorable to the Milankovitch theory, as discussed earlier. In that case, the Pacemaker results would have been equivocal, at best. The results from the shorter RC11-120 core would have been in agreement with expectations of the Milankovitch theory, but the results from the longer E49-18, located just a short distance from the RC11-120 core, would not have provided strong support for the theory. Yet, because the Pacemaker authors did not plot those data on their graph, readers of the paper (including the paper's referees) were denied the opportunity to question whether this exclusion of data was justified.

Another problem is that the data from the two cores do not seem to be publicly available, and the two surviving Pacemaker authors (at the time) did not respond to my requests for the original data.[12] Although versions of these data sets are available online, sometimes even multiple versions, these newer data sets differ from the originals used in the Pacemaker analysis. Most of these differences are due to measurement uncertainties, but in some cases data values, and even short sections of data, have been removed from more recent online versions of the data. This, of course, naturally raises the question: Which versions of the data are the real ones? I have tabulated some of these data sets online so that readers may easily compare the older and newer versions of the data.[13]

Given the great importance of the Pacemaker paper for the field of paleoclimatology, the fact that the original data used in the paper are very hard to come by seems rather odd. Because I was unable to obtain the original data used in the Pacemaker

Idealized depiction of Earth's magnetic field

paper, I reconstructed the six data sets, point by point, from Figures 2 and 3 in the Pacemaker paper in order to replicate their results.

Another serious problem is that the Pacemaker authors obtained their $\delta^{18}O$ values using the shells of planktonic foraminifera. Planktonic forams float freely in the water. The local water temperature at the time of formation of the foram shell influences $\delta^{18}O$ values within the shell. But temperatures in the water column can vary with both location and time. Water temperature can vary by as much as 16°F within a vertical band of water called the *thermocline*, and there is no way that scientists can possibly know at what depth a foram was floating while it was constructing its shell. Likewise, large eddies called *warm* and *cold core rings*, with diameters of tens to hundreds of kilometers, can have temperatures significantly different from the surrounding water, and they can persist for months to years.[14,15] This means that $\delta^{18}O$ sediment values, supposedly indicators of global climate change, can be significantly influenced by local factors.

Secular scientists usually attempt to minimize the influence of local effects by averaging the $\delta^{18}O$ signals from different sediment cores in a process called *stacking*. Since these local effects are thought to be random, this averaging process is thought to cancel out the local effects, leaving only the true global signal.

However, the Pacemaker authors did not stack their data sets in this way. Although they stuck data from one core end-to-end with data from a second core, no such averaging of the data was done. So, even if one accepts the usual uniformitarian assumptions, were the $\delta^{18}O$ values used by the Pacemaker authors really reliable indicators of global climate? However, the most serious problem with the Pacemaker results involves the timescales assigned to the two cores, particularly the longer E49-18 core. As noted earlier, secular scientists used an assumed age of 700,000 years for the most recent magnetic reversal to set up their timescales for the cores, even though they now claim that the age of this reversal is 780,000 years. One may then wonder: Is there additional evidence for the Milankovitch theory? We will discuss that question in appendix D.

References

1. Hays, J. D., J. Imbrie, and N. J. Shackleton. 1976. Variations in the Earth's Orbit: Pacemaker of the Ice Ages. *Science.* 194 (4270): 1121-1132.
2. Muller, R. A. and G. J. MacDonald. 2002. *Ice Ages and Astronomical Causes: Data, Spectral Analysis, and Mechanisms.* Chichester, UK: Praxis Publishing, 74-78.
3. Hebert, J. 2016. Should the "Pacemaker of the Ice Ages" Paper Be Retracted?—Part 1. *Answers Research Journal.* 9: 25-56.
4. Hebert, J. 2016. Should the "Pacemaker of the Ice Ages" Paper Be Retracted?—Part 2. *Answers Research Journal.* 9: 131-147.
5. Hebert, J. 2016. Should the "Pacemaker of the Ice Ages" Paper Be Retracted?—Part 3. *Answers Research Journal.* 9: 229-255.
6. Hebert, J. 2016. Milankovitch Meltdown: Toppling an Iconic Old-Earth Argument, Part 1. *Acts & Facts.* 45 (11): 10-13.
7. Hebert, J. 2016. Milankovitch Meltdown: Toppling an Iconic Old-Earth Argument, Part 2. *Acts & Facts.* 45 (12): 10-13.
8. Hebert, J. 2017. Milankovitch Meltdown: Toppling an Iconic Old-Earth Argument, Part 3. *Acts & Facts.* 46 (1): 10-13.
9. Howard, W. R. and W. L. Prell. 1992. Late Quaternary Surface Circulation of the Southern Indian Ocean and its Relationship to Orbital Variations. *Paleoceanography and Paleoclimatology.* 7 (1): 79-117.
10. Hebert, J. 2016. Should the "Pacemaker of the Ice Ages" Paper Be Retracted?—Part 3, pp. 242-243, 246-247. Pay special attention to Figures 11-13 and Figures 24-26.
11. Howard, W. R. and W. L. Prell. Late Quaternary Surface Circulation, pp. 87-91.
12. Nicholas Shackleton passed away in 2006, as did John Imbrie in 2016, so Imbrie may have been in poor health at the time. However, Jim Hays did not respond to my request for his original data either.
13. Hebert, Should the "Pacemaker of the Ice Ages" Paper Be Retracted?—Part 1, 38-56.
14. Backus, R. H. et al. 1981. Gulf Stream Cold-Core Rings: Their Physics, Chemistry, and Biology. *Science.* 212 (4499): 1091-1100.
15. Brown, O. B. et al. 1986. Gulf Stream warm rings: a statistical study of their behavior. *Deep Sea Research Part A. Oceanographic Research Papers.* 33 (11-12): 1459-1473.

D Appendix

Is There Additional Evidence for the Milankovitch Theory?

Summary: Despite the enormous number of published papers dealing with the Milankovitch theory, there appears to be little, if any, supporting evidence for it. A 1997 paper published in the journal *Paleoceanography* seems to be an effort to salvage the results from the Pacemaker paper. It attempts to provide a justification for the timing of past Ice Ages that doesn't depend upon an assumed age of 700,000 years for the most recent magnetic reversal. However, this attempt is not terribly convincing. Contrary to popular perception, secular scientists do not view radioisotope dates as absolute. Scientists often let assumptions, theory, and even politics manipulate these ages. The revision to the age of the Brunhes-Matuyama magnetic reversal is a classic example. Secular scientists used an assumed age of 700,000 years for this event to "confirm" the validity of the Milankovitch theory. This age was later revised upward to 730,000 years, based on better data. However, in the early 1990s, scientists arbitrarily revised the age upward to 780,000 years, over-ruling their own previous radioisotope age estimates. The apparent justification for this revision was that scientists were having difficulty reconciling other data with the Milankovitch theory unless the ages for the most recent magnetic reversals were changed. So, they did revise these ages, but in the process, they undermined the original argument for the theory!

In spite of the serious problems with the "Pacemaker of the Ice Ages" paper, it is tempting to think that evidence for the Milankovitch theory must still be very strong. After all, an internet search for "Milankovitch theory" yields over 100,000 results. Given the sheer number of papers and articles dealing with this subject, someone might easily think that the theory surely rests on a secure foundation. However, most of these papers simply assume the theory is correct and then use it to draw conclusions of some kind. For instance, many papers use this assumption to assign ages to chemical wiggles in seafloor sediment or ice cores (orbital tuning). Others use the Milankovitch theory to draw inferences about past or future climate change. In this appendix, I explain why additional supporting evidence for the theory is probably weak, if not nonexistent.

Why Was the Age for the M-B Reversal Revised?

Before doing so, we need to discuss why uniformitarian scientists revised the age of the Brunhes-Matuyama magnetic reversal in the first place. In 1976, the accepted age of this magnetic reversal, based on radioisotope dating of volcanic rocks, was 700,000 years. This was the age secular scientists used to set up the timescales for the deep-sea sediment cores featured in the Pacemaker paper.[1,2] However, in 1979, the age of the M-B reversal was adjusted upward slightly to 730,000 years based on new data and revised values for the potassium-argon dating constants.[3] By itself, this revision was probably not large enough to adversely affect the Pacemaker results.

However, in the early 1990s a number of scientists began advocating that the ages for the most recent magnetic reversals be revised upward, including the M-B reversal. In 1990, Nicholas Shackleton, one of the Pacemaker authors, published a paper in which he advocated that the ages for the two most recent magnetic reversals, including the M-B reversal, should be increased 5 to 7%.[4] Although this may not sound like much, a 7% increase in the age of 730,000 years for the M-B reversal raised it to 780,000 years.

Why did Shackleton make this recommendation? He was using the Milankovitch theory to assign ages to oxygen isotope wiggles within a deep-sea sediment core near the Galápagos Islands. In other words, he used the theory to orbitally tune those $\delta^{18}O$ values. Although he and his co-authors never came right out and said it, the lower ages for the most recent magnetic reversals were apparently making it difficult for them to assign ages to the $\delta^{18}O$ values that agreed with both the Milankovitch theory and the then-accepted ages for those magnetic reversals. Hence, he advocated revising the ages of those reversals upward. Likewise, a scientist named F. J. Hilgen made a similar recommendation in 1991 based on analysis of a series of Mediterranean cores.[5]

In other words, secular scientists allowed the Milankovitch theory to overrule their radioisotope age estimates. Most people are under the impression that radioisotope dating methods are objective and independent methods of dating rocks. They would be surprised to learn that secular scientists often allow outside political considerations to influence these radioisotope dating methods.

"In other words, secular scientists allowed the Milankovitch theory to overrule their radioisotope age estimates."

Although this revised age for the M-B reversal was later "confirmed" by radioisotope dating,[6] this confirmation was not all that impressive.[7] Secular scientists used a mathematical technique to pick an age for each reversal that gave the best overall fit to a large number of magnetic measurements from volcanic rocks. The original analysis resulted in an age of 730,000 years for the M-B reversal giving the best overall fit (the smallest overall inconsistency) to dozens of data points from around the world (New Mexico, California, Alaska, Australia, the Canary Islands, etc.).[8]

However, when secular scientists realized they needed a higher age of 780,000 years, they used this method again, but after including additional data from New Mexico and by allowing the algorithm to include higher ages in its "search." Not too surprisingly, these additional data points and this loosened constraint nudged the answer toward the needed higher age assignment of 780,000 years.[9] Another analysis using magnetic data from Hawaii seemed to confirm this higher age estimate.[10] It is hardly surprising that different sets of data can give different answers, so how do these new estimates actually "prove" anything? It looks like the Milankovitch-theory tail has been wagging the radioisotope-dating dog!

Reasons to Doubt Milankovitch

Now that we've explained why secular scientists made this age revision, we can ask whether or not they have additional hard evidence for the Milankovitch theory. We will look at four reasons to doubt the theory. First, even if one accepts the usual uniformitarian assumptions, confirmation of the Milankovitch theory is not easy. You need a long, undisturbed sediment core (or cores), and you need a way to independently assign ages to the chemical wiggles in that core, a way that doesn't simply *assume* that the Milankovitch theory is correct. The Pacemaker authors claimed that only two sediment cores had the necessary properties for their analysis, even though at the time hundreds of sediment cores had already been drilled.[11] So, finding cores having the right properties is difficult. And since radioisotope dating methods (with a few exceptions[12]) cannot be used directly on the sediments, then dating the chemical wiggles is not easy either. Many secular scientists saw the V28-238 western Pacific sediment core as a way of assigning ages to chemical wiggles because they had convinced themselves that those sediments were deposited at a nearly constant rate over time (see chapter 9). Since the ages at two locations in the core were thought to be known (the age at the top of the core and the age at a depth of 1,200 cm), uniformitarian scientists thought this allowed them to easily assign ages to prominent features in the

oxygen isotope wiggles in the core. The V28-238 core was so important in their thinking that some referred to it as an "Ice Age Rosetta Stone," because it seemed to provide an objective way of dating the chemical wiggles in seafloor sediments.[13]

Second, uniformitarian oceanographer Wolfgang Berger made a jaw-dropping statement about what he considered the "strongest argument yet" for the Milankovitch theory:

> In the end, the correct timescale [for the marine sediment cores] was a matter of co-ordinating isotope stratigraphy with the results from palaeomagnetism, applying the date found in basalt layers for the Matuyama-Brunhes boundary to cores with known magnetic stratigraphy (as in Shackleton & Opdyke 1973). The agreement of dating by that method and by Milankovitch tuning (urged by Shackleton et al. 1990) is the strongest argument yet for the correctness of Milankovitch theory.[14]

Remember that the Pacemaker authors obtained their age estimates from the 1973 paper by Shackleton and Opdyke, which assumed an age of 700,000 years for the Brunhes-Matuyama magnetic reversal boundary. However, the 1990 paper by Shackleton et al recommended a completely different age of 780,000 years for that same

Mechanical ice core drill

boundary. Although the Pacemaker paper did not provide error bars for the assumed age of 700,000 years for the M-B reversal, uncertainties in radioisotope ages for the M-B reversal that have been published are usually around 10,000 years—much too small to account for the apparent 80,000-year difference in those ages.[15-17]

This means that the 1973 Shackleton and Opdyke paper and the 1990 paper by Shackleton et al required two completely different age estimates for the *same* magnetic reversal. Yet, Berger says that the agreement between those two papers "is the strongest argument yet for the correctness of Milankovitch theory." If the "strongest argument yet" for the Milankovitch theory requires two completely different ages for the same magnetic reversal, then I think it is fair to say that the theory is in serious trouble!

"If the 'strongest argument yet' for the Milankovitch theory requires two completely different ages for the same magnetic reversal, then I think it is fair to say that the theory is in serious trouble!"

That brings us to a third reason to suspect that additional evidence for the Milankovitch theory is lacking. If one needs to use different ages for the same event in order to reconcile all the data with the theory, then why should anyone take the theory seriously?

Fourth, the Pacemaker results, if taken at face value, imply that the climate responds to astronomical changes so that the lengths of the climate cycles are the same as the lengths of the astronomical cycles. In other words, astronomical cycles of 41K, 23K, and 100K years result in climate cycles that are *also* 41K, 23K, and 100K years long.[18] However, one does not have to make this assumption. One could also assume that the climate cycles have lengths different from those of the astronomical cycles, and some uniformitarian scientists have argued for such alternate versions of the Milankovitch theory.[19,20] But this means that uniformitarian scientists have not settled even the most basic elements of the theory. How does one test a theory whose details are still in dispute?

Has the Pacemaker Paper Been Rescued?

However, if you were to press uniformitarian scientists on this issue, they might attempt to use a 1997 paper by paleoclimatologist Maureen Raymo to argue that the Pacemaker results are still valid.[21] Dr. Raymo used chemical wiggles ($\delta^{18}O$ values) from 11 sediment cores, all of which contained the M-B magnetic boundary, to argue for the validity of the Milankovitch theory. She used an assumed age of 780,000 years for the M-B reversal, as well as two other radioisotope age assignments, to date a number of marine isotope stage (MIS) boundaries. Her age estimates for those boundaries agreed well with the ages used in the Pacemaker paper. She did not use spectral analysis to confirm the Milankovitch theory, but there was no need for her to do so. Since the ages for those boundaries agreed well with the ages originally used in the Pacemaker paper, it was a foregone conclusion that the "new" Pacemaker results would again confirm the Milankovitch theory.[22] For this reason, this method superficially seems to salvage the results of the original Pacemaker paper, but it implicitly repudiates the method used to obtain the original age estimates for the MIS boundaries.[23]

However, Raymo's selection of sediment cores was arguably biased. There are quite a few sediment cores containing the M-B magnetic reversal that she did not use in her analysis. In fact, a paper published just one year prior to hers mentioned seven other sediment cores containing the M-B reversal. Yet, Raymo did not include these seven cores in her analysis.[24] This suggests that the cores used in Raymo's analysis may not have been a fair sampling of all the available data. This is further suggested by her statement that she used in her analysis sediment cores that "exhibit the *typical sequence of oxygen isotope stages* with no obvious hiatuses, drilling disturbance, or large swings in sedimentation rate."[25]

Obviously, one should avoid disturbed or distorted sediment cores when performing such an analysis, but this raises an extremely important question: How does one know what an undistorted "typical" oxygen isotope sequence is supposed to look like? Secular scientists think that an "ideal" undistorted oxygen isotope record should display a recognizable sawtooth pattern (Figure B.1). They think this sawtooth pattern represents repeated slow buildups of ice sheets during Ice Ages and then their rapid collapses during the intervening warm periods. So, secular scientists tend to assume that a core that doesn't have this expected sawtooth pattern has somehow been distort-

Glacier Perito Moreno National Park, Argentina

ed. But they don't really know that. Perhaps the core was undistorted but the sawtooth pattern simply wasn't present for whatever reason. As I describe in appendix B, because we don't really know how past sedimentation rates have varied over time, we can't really know for certain what an undistorted oxygen isotope record is supposed to look like. So, by restricting her analysis to sediment cores having the expected $\delta^{18}O$ pattern, Dr. Raymo may have unconsciously biased her results.

The most striking thing about her analysis was her exclusion of data from the V28-238 core, even though this was the core used by Shackleton and Opdyke to obtain their original age estimates for the MIS boundaries. Raymo said she excluded these cores because they had already been orbitally tuned to the so-called SPECMAP timescale, which she was attempting to confirm. However, that sounds like an excuse for exclusion rather than a reason. Since secular scientists still presumably had the V28-238 $\delta^{18}O$ data as a function of core depth, it should have been a trivially simple matter to un-tune those $\delta^{18}O$ values and to treat them like oxygen isotope data from any other core. Moreover, since the V28-238 core was supposedly characterized by a nearly constant sedimentation rate, it should definitely have been included in the analysis, and it should have been given greater weight than the data from the other cores.

Remember that secular scientists used to call the oxygen isotope data from the V28-238 core an Ice Age Rosetta Stone because of that core's supposed importance.[26] Of course, they said that when the core was giving them age estimates they wanted to hear. Now that the revised age estimate for the M-B boundary has messed up those original results, the V28-238 core has apparently outlived its usefulness. Secular scientists have abandoned the V28-238 core, having become enamored with other sediment cores that now give them the right answers. Now the V28-238 sediment core has been quietly demoted to just another run-of-the-mill sediment core among thousands of other such cores!

Are Raymo's results convincing? I don't think so. In order to accept her results at face value, one has to assume that Shackleton and Opdyke somehow, through sheer luck, obtained reasonably accurate age estimates for the MIS boundaries, even though their original methodology, by their own reckoning, was completely wrong. This sounds suspiciously like the high school student who complains that the teacher should not have marked his homework wrong, because he got the "right" answer, despite incorrect reasoning!

“To the best of my knowledge, secular scientists have never candidly acknowledged this issue, even after creationists pointed it out in 2016. Their failure to do so calls their objectivity on this subject into serious question.”

Remember also that radioisotope dating supposedly confirmed the age of the M-B reversal boundary in 1992.[27] This means there was a five-year period between that 1992 “confirmation” and the publication of Raymo’s paper in 1997 during which there may have been no real evidence for the Milankovitch theory—even by secular reckoning. Then why didn’t paleoclimatologists sometime between 1992 and 1997 inform both the larger scientific community and the general public of this serious problem? To the best of my knowledge, secular scientists have never candidly acknowledged this issue, even after creationists pointed it out in 2016.[28] Their failure to do so calls their objectivity on this subject into serious question.

In any case, secular scientists are being arbitrary and capricious in their handling of the seafloor sediment data. One gets the distinct impression that they are simply unwilling to allow the Milankovitch theory to be falsified by contrary evidence. When a particular method (such as the spectral analysis used in the Pacemaker paper) seems to provide evidence for the Milankovitch theory, all educated people are expected to accept and believe those results. In fact, if you don’t accept those results, you’re ridiculed as scientifically ignorant. Yet, when that exact same method provides evidence against the theory (due to an age revision that the secular scientists themselves made), then apparently we aren’t supposed to know about that. And even if we *do* learn about it, we’re apparently supposed to not pay attention to it!

A Geochronological Train Wreck

As discussed in chapters 18 and 19, this has huge implications for the global warming debate. But it also has enormous implications for the debate over the earth’s age. One of the frequent objections to biblical creation is the supposed agreement between different dating methods. Given the apparent agreement between so many different and supposedly independent dating methods, how can any reasonable person doubt that the earth is billions of years old?

However, the age revision to the M-B magnetic reversal shows that the methods do *not* all agree. Sometimes, serious contradictions are lurking beneath the surface. Unfortunately, these contradictions are usually buried in the technical literature where very few people see them.

However, the wiggle-matching process discussed in chapter 9 also contributes to this apparent agreement. Remember that secular scientists think prominent oxygen isotope features in different sediment cores should theoretically be the same age. This means that if they can somehow assign an age to a $\delta^{18}O$ feature in one sediment core, that age assignment can presumably be transferred to a similar feature in another sediment core.

Secular scientists routinely use this wiggle-matching process to assign ages to the chemical wiggles in sediment and ice cores (Figure D.1). In fact, it is not at all unusual

for ages assigned to wiggles in one particular core to be "tied" to ages assigned to multiple other cores. For instance, ages assigned to one particular sediment core off the coast of New Zealand (designated as "A" in Figure D.2) were tied to ages assigned to ice cores from both Greenland and Antarctica, as well as sediment cores off the coasts of Portugal, Iceland, and the western coast of Ecuador.[29,30] Nearly all those ages came from the Milankovitch theory. But what happens if the evidence for the Milankovitch theory is weak or, worse yet, nonexistent? Then those age assignments are all suspect (Figure D.3), even for evolutionists who believe in an old earth.

So, the apparent agreement between different dating methods may at first seem to be a strength for the secular worldview, but it is also one of its greatest weaknesses. That is because much of this apparent agreement is coming from wiggle matching. But wiggle matching doesn't *prove* that chemical wiggles in one core are the same age as the corresponding wiggles in another core, it simply *assumes* that this is the case. Remember, these age assignments are almost never based on radioisotope dating. They are based on the assumption that the Milankovitch theory is correct. So, in order to topple all these age assignments, you don't have to refute ages assigned to every single sediment and ice core. Because nearly all these ages are coming from the Milankovitch theory, you simply have to show that the evidence for the Milankovitch theory itself is lacking, as I have attempted to do here. Without firm evidence for the theory, these age assignments logically come crashing down like a train of falling dominos!

Lest you think I am exaggerating, consider the following comments from a 1997 *Science News* article. This article was describing criticisms of the Milankovitch theory by Berkeley physicist Richard Muller at a meeting of paleoclimate researchers in 1996. Muller was not advocating a total abandonment of the theory but was suggesting that scientists had misidentified the true cause of the 100,000-year climate cycle that seemed apparent in seafloor sediments. The article said:

> [Richard] Muller scored the most points at the meeting when he attacked a standard technique, called tuning, that oceanographers used for dating layers in sediment cores. The task of dating these strata is difficult because sediments may accumulate more quickly during some eras and more slowly in others. To tell the age of layers between known benchmarks, researchers often use the Milankovitch orbital cycles to

> tune the sediment record: They assume that ice volume should vary with the orbital cycles, then line up the wiggles in the sediment record with ups and downs in the astronomical record.
>
> "This whole tuning procedure, which is used extensively, has elements of circular reasoning in it," says Muller. He argues that tuning can artificially make the sediment record support the Milankovitch theory. Muller's criticisms hit home with many researchers. "He scared the [expletive] out of them, and they deserved it," says [climate scientist W. S.] Broecker.[31]

Why were these researchers so frightened? The answer is obvious. If hard evidence for the Milankovitch theory was lacking, then all their conclusions from 20 years of research were immediately suspect. In fact, there is another interesting aspect to this 1996 meeting. Apparently, Muller's criticisms at this meeting prompted Maureen Raymo to publish her 1997 paper. Immediately after the above statement, the *Science News* article continued:

> Oceanographers soon rose to the challenge. In the August *Paleoceanography*, Maureen E. Raymo of Massechusetts Institute of Technology presents an untuned sediment record that corroborates the ice age dates determined by dating.[32]

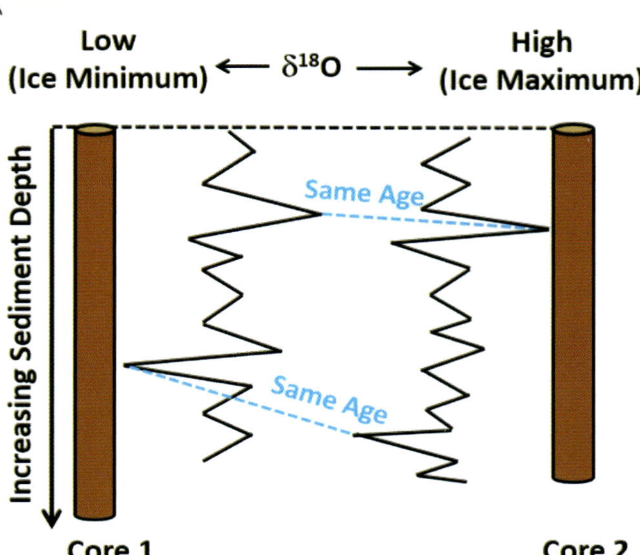

Figure D.1. Secular scientists use "wiggle matching" to transfer ages assigned to chemical wiggles in one sediment core to supposedly corresponding chemical wiggles in other sediment cores

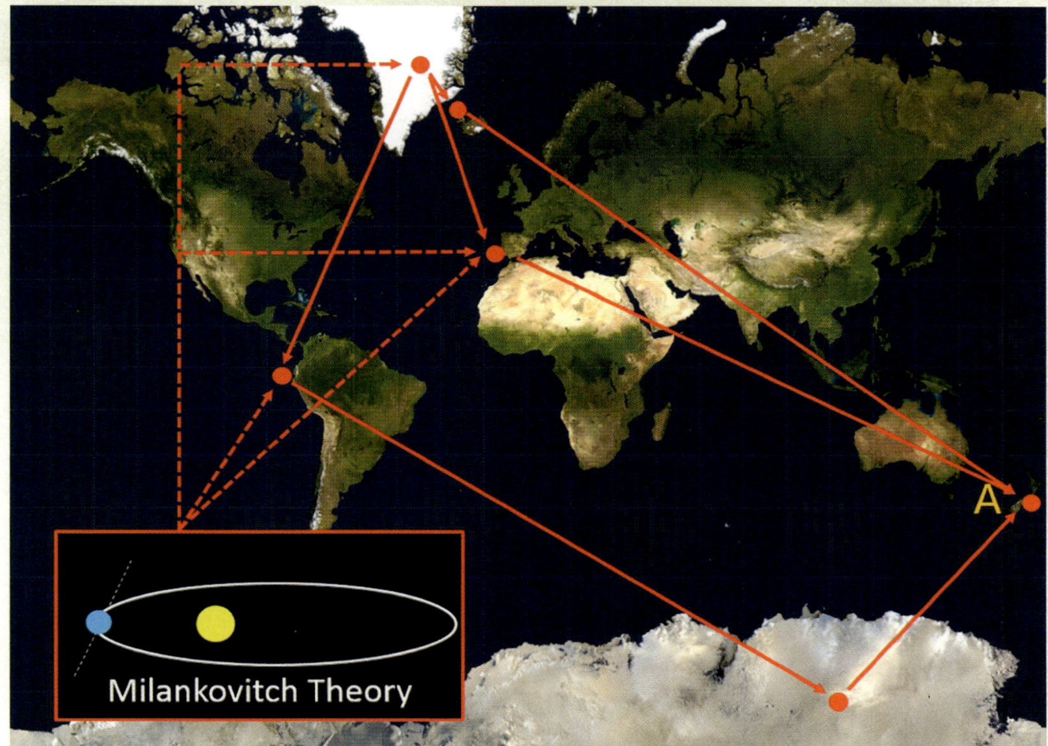

Figure D.2. Ages assigned to the deep-sea sediment core MD97-2120 off the coast of New Zealand were themselves tied back to ages assigned to other sediment and ice cores, most of which were ultimately tied back to the Milankovitch Ice Age theory

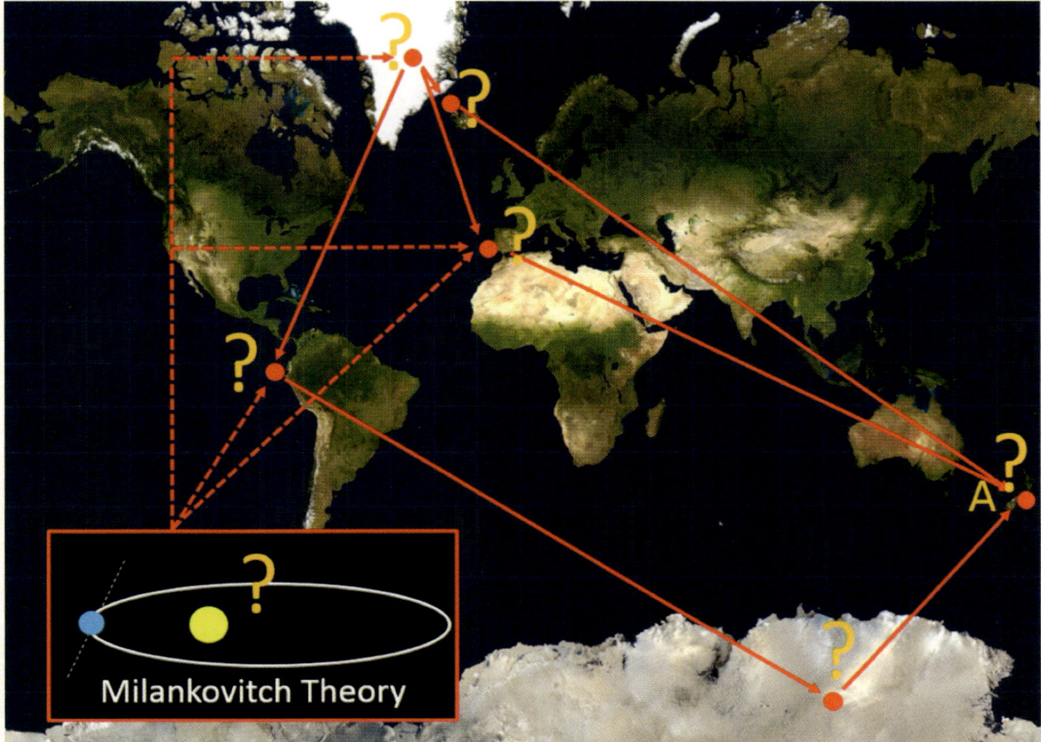

Figure D.3. Without strong evidence for the Milankovitch theory, these age assignments are in doubt, even by secular, old-earth reckoning

Yes, this August *Paleoceanography* paper is the one I described earlier. I strongly suspected that Dr. Raymo's paper was an attempt to provide objective evidence for the Milankovitch theory. After reading this, I *know* that it was. Apparently, someone said something at this 1996 meeting that so alarmed secular paleoclimatologists that Dr. Raymo felt the need to quickly (in about a year!) publish a paper that ostensibly confirmed the Milankovitch theory. But didn't the Pacemaker of the Ice Ages paper already confirm the theory? Why the need for an alternate proof?

For the reasons above, I was already very skeptical of this alleged confirmation of the theory. After learning that Raymo published it just a year or so after this meeting, I am even more suspicious. Remember that secular scientists have long sought to confirm the Milankovitch theory. At least one previous attempted confirmation, prior to the Pacemaker paper, had already been shot down by secular scientists themselves.[33,34] Then it took the Pacemaker authors another 10 years to come up with another justification for the theory. I find it rather amazing (and awfully convenient) that Dr. Raymo was able to so quickly provide this needed confirmation of the Milankovitch theory. That sounds like an academic rush job.

"For the reasons above, I was already very skeptical of this alleged confirmation of the theory. After learning that Raymo published it just a year or so after this meeting, I am even more suspicious."

By the way, that Dr. Raymo apparently felt the need to publish this paper is additional evidence that secular scientists have little, if any, hard evidence for the Milankovitch theory, even by their own reckoning. The researchers at the 1996 meeting were surely aware of the existence of the Pacemaker paper. So, why did Dr. Raymo feel the need to quickly get this paper published? And why did she (apparently) publish it in response to the charge that orbital tuning was an exercise in circular reasoning? That Raymo felt the need to do so strongly suggests that the scientists at this 1996 meeting did *not* consider the 20-year-old Pacemaker paper to be a strong argument for the Milankovitch theory.

I can't help but wonder: Did someone at this 1996 meeting point out the problem with the age of the M-B reversal 20 years before I did? Regardless of the answer to that question, it's obvious that by 1996 secular scientists were uncomfortable, for whatever reason, basing their entire argument for the Milankovitch theory on just the Pacemaker paper results.

I am not saying that Dr. Raymo or others were being intentionally dishonest, but I am suggesting that their apparent bias in favor of the Milankovitch theory may have prevented them from seeing the circular nature of the argument they were making.

In fact, to say that the orbital tuning exercise merely contained elements of circular reasoning is a gross understatement. Imagine that the orbital tuning exercise is a house that has been erected on the foundation of the Milankovitch theory. The Milankovitch theory, in turn, rests on the foundation of the 1976 Pacemaker paper. But the 1976 Pacemaker paper rests on still *another* foundation, the results of the 1973 paper by Shackleton and Opdyke. But when that lower foundation crumbled (Figures D.4 and D.5), there was no longer a logical basis for thinking that the Milankovitch theory

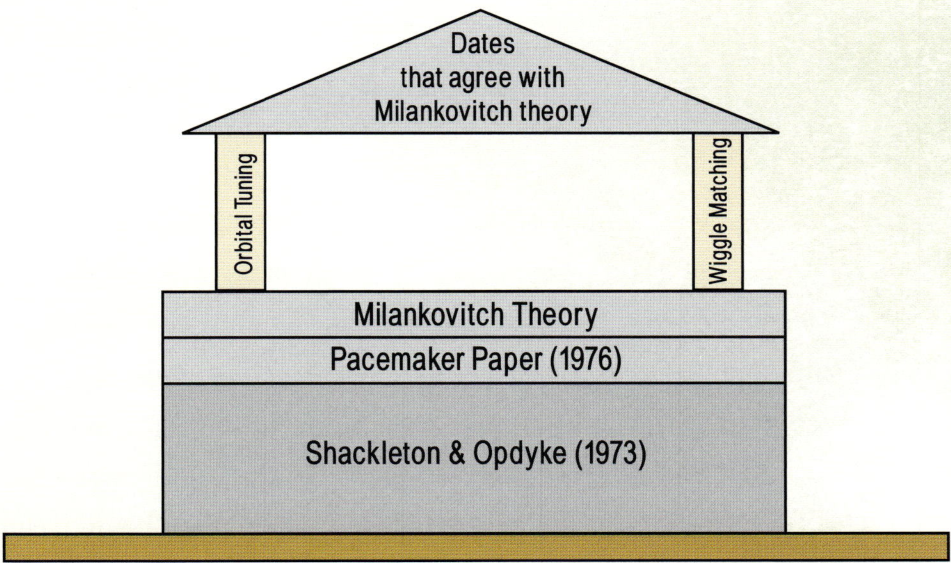

Figure D.4. Secular scientists have made extensive use of orbital tuning to construct their secular geochronology. However, the validity of orbital tuning rests on the validity of the Milankovitch theory. That, in turn, rests on the validity of the 1976 "Pacemaker of the Ice Ages" paper, which itself in turn rested on the validity of the 1973 paper by Shackleton and Opdyke. Yet, the age revision to the M-B magnetic reversal invalidated both the 1973 paper and, by implication, the 1976 Pacemaker paper.

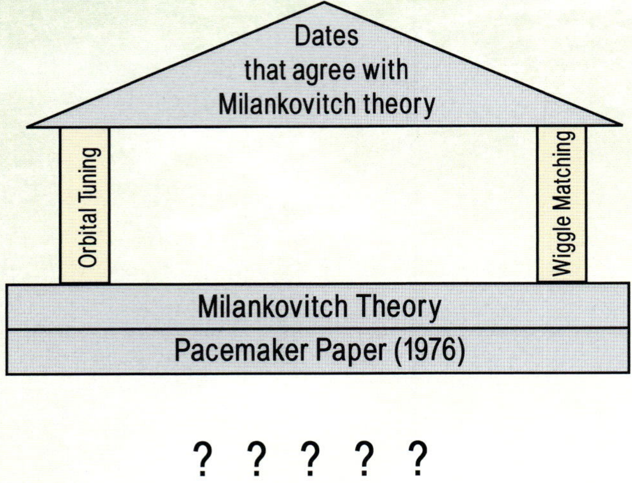

Figure D.5. Without a firm logical foundation, the entire orbital tuning exercise is, logically speaking, floating in midair, with nothing underneath to justify it

was correct. I strongly suspect that the Raymo paper (Figure D.6) was an attempt to provide an alternate justification for the age assignments used in the Pacemaker paper. I would argue that this attempted justification is not very convincing.

As suggested above, the absence of strong evidence for the Milankovitch theory is devastating to secular age assignments. A 1996 news article describes Richard Muller's alternate explanation for the 100,000-year cycle that is apparent (after making the usual uniformitarian assumptions) in deep-sea sediments:

> It sounds like a scientific success story. But this kind of success carries within it the seed of its own problems. New ideas are particularly hard to accept when they overthrow the fundamental assumption on which most of the work in the field is based. As Muller, who has a nice way with an aphorism, wrote in [his book] *Nemesis*, "Old theories never die, only old theorists."
>
> So far, Muller and [co-researcher Gordon] MacDonald have been unable to publish their full paper despite their considerable credentials. It's been rejected by *Science*. It's been rejected by *Nature* three times though the editors did request and publish a shorter version summarizing their findings in November of 2018. Why? Muller pulls open a long [sic] file drawer, crammed with papers. "Here it is. Essentially

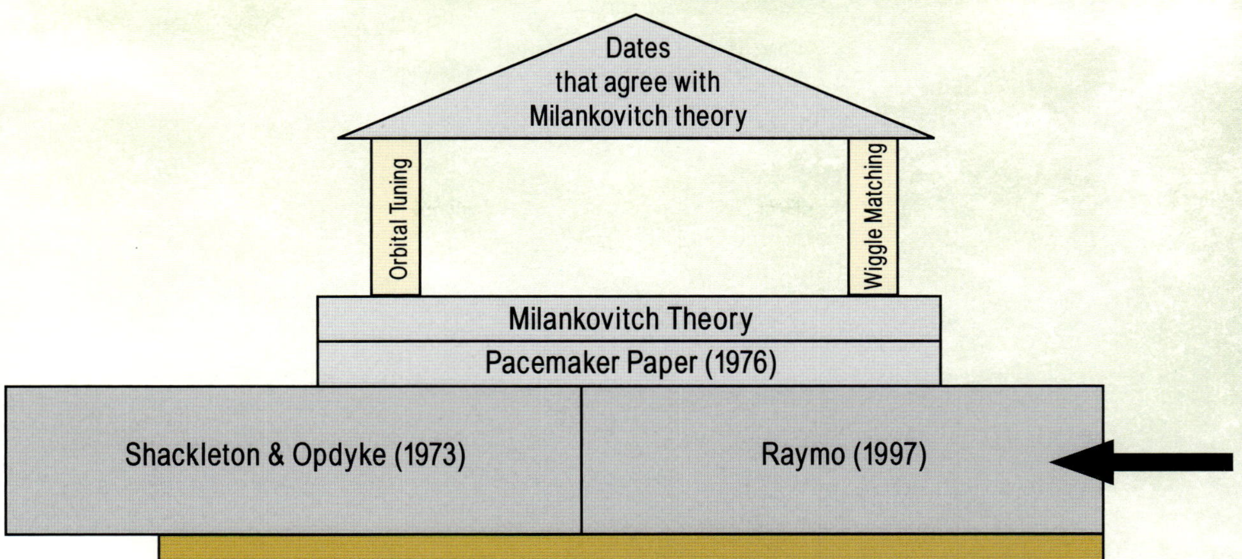

Figure D.6. Maureen Raymo's 1997 paper appears to be an attempt to provide an alternate justification for the age assignments originally obtained from the 1973 Shackleton and Opdyke paper that the Pacemaker authors used in their analysis. This would enable secular scientists to claim that the Pacemaker results are still valid, despite their age revision to the M-B magnetic reversal boundary.

> everything that's been published for the last twenty years assumes the Milankovitch model. I think it's very hard for people in this field, and all the referees to whom our paper has been submitted are working in this field, to accept our paper. They'd have to say that most of their own work for the past twenty years is fundamentally flawed."[35]

Of course, that article was published in 1996. Today, even more papers have since been published, papers that assume the Milankovitch theory to be true and use that assumption to draw conclusions about climate change or to assign ages to ice and sediment cores. What happens to all these results if evidence for the theory is weak or nonexistent?

By the way, the above *Science News* article clearly demonstrates that the sheer number of Milankovitch-based papers tells us nothing about the actual strength of the evidence for the theory. Twenty years after the publication of the Pacemaker paper, a long file drawer of Milankovitch papers, to quote the *Science News* article, had already been published. Dr. Raymo and other paleoceanographers were obviously familiar with these papers. Yet, despite the large number of Milankovitch papers published during the previous two decades, she apparently did not think those papers made a sufficiently strong argument for the theory. Otherwise, she would not have felt compelled to come up with an alternate argument for the ages assigned to the Ice Ages. Hence, it is a huge mistake to naively assume that the theory is strong based just on the sheer number of published papers.

By the way, secular scientists routinely use the Milankovitch theory to calibrate a particular radioisotope dating method called argon-argon dating.[36,37] When push comes to shove, they trust the Milankovitch theory more than they trust radioisotope dating. At the risk of sounding like a broken record, if evidence for the Milankovitch theory is weak, then those ages are in doubt too—even by secular reckoning.

And it's not just these age assignments that are in question. The conclusions of *any* published paper whose results depended in any way upon the assumption that the Milankovitch theory is correct are also in question. Probably thousands of such papers exist. This truly is a train wreck for uniformitarian scientists. And that's no hype.

References

1. Shackleton, N. J. and N. D. Opdyke. 1973. Oxygen Isotope and Palaeomagnetic Stratigraphy of Equatorial Pacific Core V28-238: Oxygen Isotope Temperatures and Ice Volumes on a 10^5 Year and 10^6 Year Scale. *Quaternary Research.* 3 (1): 39-55.
2. Hays, J. D., J. Imbrie, and N. J. Shackleton. 1976. Variations in the Earth's Orbit: Pacemaker of the Ice Ages. *Science.* 194 (4270): 1124.
3. Mankinen, E. A. and G. B. Dalrymple. 1979. Revised Geomagnetic Polarity Time Scale for the Interval 0-5 m.y. B.P. *Journal of Geophysical Research: Solid Earth.* 84 (B2): 615- 626.
4. Shackleton, N. J., A. Berger, and W. R. Peltier. 1990. An Alternative Astronomical Calibration of the Lower Pleistocene Timescale Based on ODP Site 677. *Transactions of the Royal Society of Edinburgh: Earth Sciences.* 81 (4): 251-261.
5. Hilgen, F. J. 1991. Astronomical calibration of Gauss to Matuyama sapropels in the Mediterranean and implication for the Geomagnetic Polarity Time Scale. *Earth and Planetary Science Letters.* 104 (2-4): 226-244.
6. Herbert, T. D. 2010. Paleoceanography: Orbitally Tuned Timescales. In *Climate and Oceans.* J. H. Steele, ed. Amsterdam, Netherlands: Academic Press, 374.
7. Hebert, J. 2019. Have uniformitarians rescued the 'Pacemaker of the Ice Ages' paper? *Journal of Creation.* 33 (1): 102-109.
8. Mankinen and Dalrymple, Revised Geomagnetic Polarity Time Scale for the Interval 0-5 m.y. B.P.
9. Spell, T. L. and I. McDougall. 1992. Revisions to the Age of the Brunhes-Matuyama Boundary and the Pleistocene Geomagnetic Polarity Timescale. *Geophysical Research Letters.* 19 (12): 1181-1184.
10. Baksi, A. K. et al. 1992. $^{40}Ar/^{39}Ar$ Dating of the Brunhes-Matuyama Geomagnetic Field Reversal. *Science.* 256: 356-357.
11. Hays et al, Variations in the Earth's Orbit: Pacemaker of the Ice Ages, 1122.
12. As a general rule, radioisotope dating methods that can theoretically be used to date seafloor sediments, such as radiocarbon and uranium-series disequilibrium dating, can only be used to date sediments that are relatively young by uniformitarian reckoning. Hence, they still need other methods to date the deeper sediments.
13. Woodward, J. 2014. *The Ice Age: A Very Short Introduction.* Oxford: Oxford University Press, 97.

14. Berger, W. H. 2013. On the Beginnings of Palaeoceanography: Foraminifera, Pioneers, and the *Albatross* Expedition. In *Landmarks in Foraminiferal Micropalaeontology: History and Development.* A. J. Bowden, F. J. Gregory, and A. S. Henderson, eds. London: Geological Society Publishing House, 169.
15. Baksi et al, ^{40}Ar/^{39}Ar Dating of the Brunhes-Matuyama Geomagnetic Field Reversal.
16. Imbrie, J. et al. 1984. The Orbital Theory of Pleistocene Climate: Support from a Revised Chronology of the Marine δ^{18}O Record. In *Milankovitch and Climate, Part 1.* A. Berger et al, eds. Dordrecht, Netherlands: D. Reidel Publishing: 269-305. Pay particular attention to page 285.
17. Mankinen and Dalrymple, Revised Geomagnetic Polarity Time Scale for the Interval 0-5 m.y. B.P., 623.
18. Actually, a 100,000-year climate cycle is a problem for the astronomical theory because there is no simple 100,000-year astronomical cycle. Rather, there are two closely spaced 95K and 125K-year astronomical cycles, but this couplet of cycles does not usually show up in high resolution data sets. See Muller, R. A. and G. J. MacDonald. 2002. *Ice Ages and Astronomical Causes: Data, Spectral Analysis and Mechanisms.* Chichester, UK: Praxis Publishing, 13.
19. Rial, J. A. and C. A. Anaclerio. 2000. Understanding nonlinear responses of the climate system to orbital forcing. *Quaternary Science Reviews.* 19 (17-18): 1709-1722.
20. Lourens. L. J. et al. 2010. Linear and non-linear response of late Neogene glacial cycles to obliquity forcing and implications for the Milankovitch theory. *Quaternary Science Reviews.* 29 (1-2): 352-365.
21. Raymo, M. E. 1997. The timing of major climate terminations. *Paleoceanography and Paleoclimatology.* 12 (4): 577–585.
22. Specifically, Raymo claimed to confirm the SPECMAP timescale for deep-sea sediments, which was orbitally tuned. Although secular scientists now agree that the older part of the SPECMAP timescale was in error, they still think that the most recent part (the part that Raymo ostensibly confirmed) is accurate.
23. Hebert, Have uniformitarians rescued the 'Pacemaker of the Ice Ages' paper?
24. Tauxe, L. et al. 1996. Astronomical calibration of the Matuyama-Brunhes boundary: consequences for magnetic remanence acquisition in marine carbonates and the Asian loess sequences. *Earth and Planetary Science Letters.* 140: 133–146.
25. Raymo, The timing of major climate terminations, 578. Emphasis added.
26. Woodward, *The Ice Age: A Very Short Introduction*, 97.
27. Spell and McDougall, Revisions to the Age of the Brunhes-Matuyama Boundary.
28. Hebert, J. 2016. Should the "Pacemaker of the Ice Ages" Paper Be Retracted?—Part 1. *Answers Research Journal.* 9: 25-56.
29. Hebert, J. 2015. The Dating "Pedigree" of Seafloor Sediment Core MD97-2120: A Case Study. *Creation Research Society Quarterly.* 51 (3): 152-164.
30. Hebert, J. 2016. Deep Core Dating and Circular Reasoning. *Acts & Facts.* 45 (3): 9-11.
31. Monastersky, R. 1997. The Big Chill: Does dust drive Earth's ice ages? *Science News.* 152 (14): 220.
32. Ibid.
33. Emiliani, C. 1966. Isotopic paleotemperatures. *Science.* 154 (3751): 851-857.
34. Oard, M. J. 1984. Ice Ages: The Mystery Solved? Part II: The Manipulation of Deep-Sea Cores. *Creation Research Society Quarterly.* 21 (3): 125-137.
35. Hurwitt, R. Get Out Your Mittens: Richard Muller thinks he knows what causes the Earth's ice ages to ebb and flow. But is anybody listening? *East Bay Express.* Posted on muller.lbl.gov July 19, 1996, accessed August 13, 2019.
36. K/Ar and ^{40}Ar/^{39}Ar Methods: The ^{40}Ar/^{39}Ar Dating technique. New Mexico Bureau of Geology and Mineral Resources: New Mexico Geochronology Research Laboratory. Posted on geoinfo.nmt.edu, accessed November 28, 2018.
37. Hebert, J. 2014. Circular Reasoning in the Dating of Deep Seafloor Sediments and Ice Cores: The Orbital Tuning Method. *Answers Research Journal.* 7: 297-309.

IMAGE CREDITS

t: top; b: bottom; l: left; r: right; m: middle

Acharya, Sanjay (via Wikipedia): 46

Baumgardner, John: 56

Bigstockphoto: 17, 19, 20, 22, 23, 24, 26-27, 29t, 35, 36-37, 39, 40-41, 48tr, 50-52, 54, 60, 64, 66, 71, 73, 74, 78, 80, 83, 84, 93, 94-95, 100-101, 102, 117, 121, 122-123b, 128-129, 131, 140, 142, 145, 147, 148, 165, 172, 175, 179, 181, 186, 189, 192, 195, 203, 204, 207, 211, 213, 217, 218, 223, 224, 228, 230-231, 246-247, 255, 265, 268, 271, 273

Brewer, Wes: 122m

Creation Science Fellowship: 57, 134

Doré, Gustav / Public Domain: 135, 136, 155t, 175

Evanson, Tim (via Wikipedia): 156

Hathaway, David/Marshall Space Flight Center: 235

Hebert, Jake: 63, 107t, 108-109, 111b, 112-113, 114, 116, 125t (after original by Michael Oard), 167-168, 170-171, 173, 183, 200, 215, 237-240, 242, 252-253, 258, 261 (drawing over map), 266, 282, 283 (drawing over map), 285-287

Humphreys, Russell: 62

ICR: 44, 125b, 132

IPCC: 206

iStockPhoto: 150-151, 256-257, 277, 280, 288-289

Juby, Ian (ianjuby.org): 201

Lan Nguyen, Marie, CCA: 48tl

Metodicar (via Wikipedia): 221

NASA: 31, 33 (Earth, graphic based on figure by Dr. Roy Spencer), 76, 90, 97, 98, 99, 111t, 160b, 184, 196, 209, 210, 233, 234, 243, 262

NASA/EOS: 205

NASA/GSFC: 126, 160t

NASA/JPL: 88, 159

NASA, Solar Dynamics Observatory: 89

NOAA: 30, 59, 92

NPS: 43tm, 45

NSF: 106, 274

Oard, Michael J.: 120, 138, 162-163

PLoS: 48b

Public domain: 91, 96, 155b, 161

Rhode, Robert A. (via Wikipedia): 216, 249

Sakaguchi, Arito and IODP.TAMU: 105

Thomas, Brian: 43b, 47

USGS: 29b, 55, 61, 107b, 110, 124, 177, 199, 251

Vardiman, Larry: 122t

Wikimedia Commons: 169, 261 (map), 283 (maps)

Windsor, Susan: 137 (after image by *Encyclopaedia Britannica*)

Wyke, Thomas / Public Domain: 245

INDEX

ABOUT THE AUTHOR

Dr. Jake Hebert earned a master's degree in physics in 1999 from Texas A&M University and a Ph.D. in physics in 2011 from the University of Texas at Dallas, where he did cutting-edge research on the connection between cosmic rays, solar activity, and weather and climate. He has taught at both the high school and university levels. He joined ICR in 2011 as a research associate and is currently researching the post-Flood Ice Age, the dating methods used to assign ages to the seafloor sediments and deep ice cores, and cosmology, among other research endeavors. He is the author of *The Ice Age and the Flood: Does Science Really Show Millions of Years?* and *The Climate Change Conflict: Keeping Cool over Global Warming*. He is also a contributor to *Guide to Creation Basics, Creation Basics & Beyond,* and *Guide to the Universe.*

ABOUT THE INSTITUTE FOR CREATION RESEARCH

After 50 years of ministry, the Institute for Creation Research remains a leader in scientific research within the context of biblical creation. Founded by Dr. Henry Morris in 1970, ICR exists to conduct scientific research within the realms of origins and Earth history and then to educate the public both formally and informally through professional training programs, through conferences and seminars around the country, and through books, magazines, and media presentations. ICR was established for three main purposes:

Dr. Henry Morris

Research. ICR conducts laboratory, field, theoretical, and library research on projects that seek to understand the science of origins and Earth history. ICR scientists have conducted multi-year research projects at key locations such as Grand Canyon, Mount St. Helens, Yosemite Valley, Santa Cruz River Valley in Argentina, and on vital issues like Radioisotopes and the Age of the Earth (RATE), Flood-Activated Sedimentation and Tectonics (FAST), the human genome, soft tissue in fossils, and other topics related to geology, genetics, astro/geophysics, paleoclimatology, paleobiochemistry, and much more.

Education. ICR offers formal courses of instruction and conducts seminars and workshops, as well as other means of instruction. With 40 years' experience in education, first through our California-based science education program (1981–2010) and now with programs offered through the School of Biblical Apologetics, ICR trains men and women to do real-world apologetics with a foundation of biblical authority and creation science. ICR's online programs include a one-year, non-degree training program for professionals called the Creationist Worldview and an Origins Matter Short Course series. Additionally, ICR scientists and staff speak to numerous groups each year through seminars and conferences, as well as offering live science presentations at the ICR Discovery Center for Science & Earth History.

Communication. ICR produces books, videos, periodicals, and other media for communicating the evidence and information related to its research and education. ICR's central publication is *Acts & Facts*, a free full-color monthly magazine with a readership of over 250,000, providing articles relevant to science, apologetics, education, and worldview issues. ICR also publishes the daily devotional *Days of Praise* with over 500,000 readers worldwide. Our website at ICR.org features regular and relevant creation science updates. The three radio programs produced by ICR can be heard on outlets around the world, and we make our materials available through multiple social media outlets.

Headquartered in Dallas, Texas, ICR's latest outreach is the ICR Discovery Center for Science & Earth History, with cutting-edge exhibits, planetarium shows, and live science presentations. The Institute for Creation Research continues to expand its work and influence, endeavoring to encourage Christians with the wonders of God's creation.

P. O. Box 59029
Dallas, Texas 75229
ICR.org

800.337.0375 (main) | 800.628.7640 (customer service)

RESOURCES FROM ICR

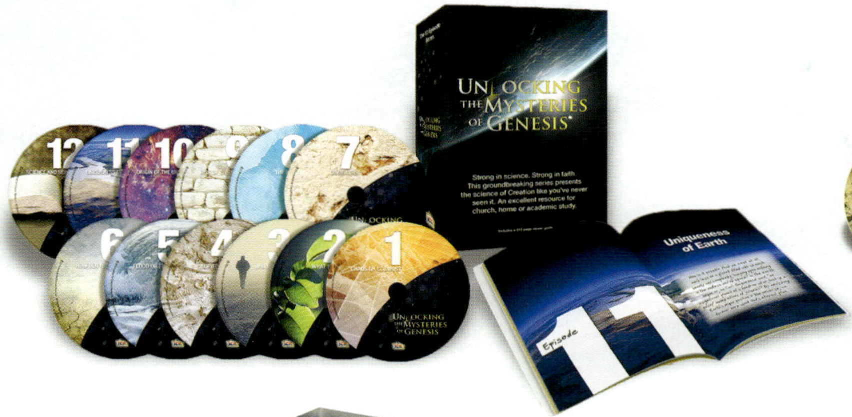

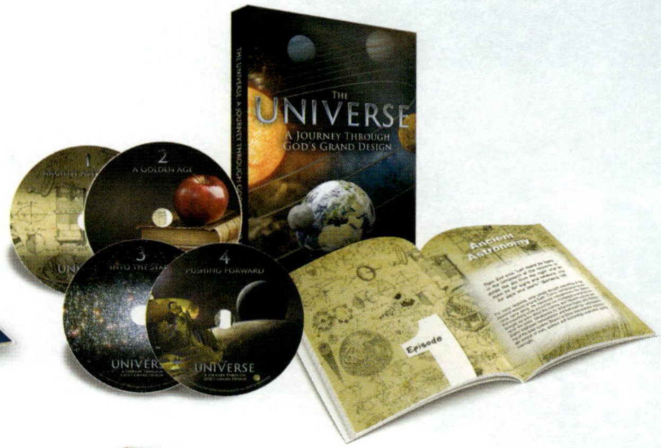

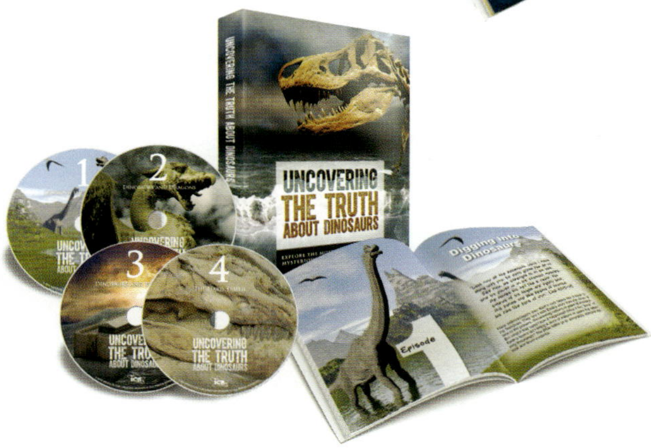

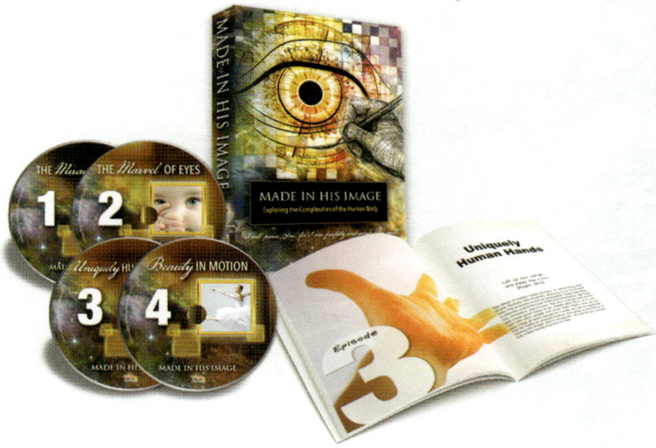

Creation or evolution? This debate is one of the most vital issues of our time. ICR's original DVD series present the evidence that confirms the biblical account of creation and provide defensible answers to questions of faith and science. Ideal for group study, these compelling and engaging presentations demonstrate that not only does the scientific evidence *not* support evolution, it strongly affirms the accuracy and authority of God's Word.

Find out more about these DVD series and other resources at **ICR.org/store**.

ICR.org